PERSHING

A History of

the Medium Tank T20 Series

by

R. P. Hunnicutt

Line Drawings
by
D. P. Dyer

Color Drawings
by
Uwe Feist

Foreword
by
Colonel Robert J. Icks, AUS-Retired

E P B M

ECHO POINT BOOKS & MEDIA, LLC

Published by Echo Point Books & Media
Brattleboro, Vermont
www.EchoPointBooks.com

Copyright © 1996, 2015 R. P. Hunnicutt
ISBN: 978-1-62654-167-2

Cover image by Uwe Feist

Cover design by Adrienne Núñez,
Echo Point Books & Media

Editorial and proofreading assistance by Christine Schultz,
Echo Point Books & Media

Printed and bound in the United States of America

CONTENTS

ACKNOWLEDGEMENTS

The material gathered into this book came from so many sources that it would be impossible to name them all. However, particular thanks must go to Colonel Robert J. Icks as well as Colonel G. B. Jarrett and the the staff of the Ordnance Museum at Aberdeen Proving Ground. Both located invaluable material. Mrs. Forst and her group at the APG Technical Library found most of the test reports on the vehicles. Carl Rasmussen, Daniel F. Smith, Robert Thibodeau, and William Russell of the Army Tank Automotive Command at Detroit Arsenal provided most of the background for the development history. This was also aided by William Tauss, who was the wartime project engineer on the M26. Description of the Pershing's introduction to combat would have been impossible without the detailed notes taken on the spot by Lt. Colonel Elmer Gray during the Zebra mission. His previously unpublished photographs were used to illustrate that section. The detailed photographs of the Pershing production came from GM Photographic by way of Dale McCormick at General Motors. Mrs. Traxler of the Army Photographic Agency located most of the Korean photos in addition to those of the Remagen crossing. Judge John Grimball, who as a young lieutenant commanded the first Pershing platoon in action, corrected much of the information on the Remagen attack. Colonel Peter Hordern was most helpful in making available the resources of the Royal Armoured Corps Tank Museum at Bovington Camp, England. Other contributors included George Schneider of the Patton Museum at Fort Knox, Joseph Avery of the National Archives, Brigadier General Joseph Colby, 1st Lt. Dwight McLemore, Paul van Thielen, Peter Chamberlain, John Kennon, L. R. "Slim" Price, and Johnny Day.

Uwe Feist and Phil Dyer labored many hours to achieve the maximum accuracy possible in their drawings. Both discovered and called the author's attention to some previously undetected variations among the experimental vehicles. Finally, I must thank my wife Susan for the long hours in typing and retyping the manuscript, despite a complete lack of interest in the subject material. All of these contributions were essential to the books completion.

FOREWORD

Reading this history of the development of the Pershing was a most pleasant experience. Most of the individuals named were familiar, including that of Lt. Colonel Elmer Gray who went through the Fort Custer Separation Center with me late in 1945.

Although a transfer from Aberdeen Proving Ground to the Office Chief of Ordnance — Detroit in February 1943 took me away from direct knowledge of the T20 series development, friends kept me informed. The first T26E1 was undergoing tests at Aberdeen not long after the all weather Munson Test Course was completed. Knowing that the course, named after a fine young officer, First Lieutenant Max Munson, who had been killed during a vehicle test about two years earlier, was designed by me with the assistance of my engineering officer, First Lieutenant William Brosha, my successor, Colonel Edward E. Gray, invited me to Aberdeen in May 1944 to see the finished course and to see the T26E1 in operation.

After touring the course in a jeep with Colonel Gray, Captain Gifford Griffin took me for a ride in the T26E1 over the varied terrain of the Churchville Test Course, which my staff and I also had laid out during my tour of duty at Aberdeen. It was an impressive tank. Compared with the M4, the ride was firm and smooth. The engine power was adequate and the transmission smooth.

But impressive as the ride was, it was not as impressive as its main armament. I had fired tank weapons many times, but never any of this caliber. Bore sighted by Johnny Day and "Slim" Price, both Automotive Division armament specialists, the 90 mm rounds hit exactly where they were aimed and gave one confidence that we soon would have a tank equal to anything the Germans had to offer.

That they were late in arriving in the combat zones can be blamed on a few Ground Forces and Armored Board officers. Most such officers were dedicated, but a few had closed minds and lacked imagination. Unfortunately, those were the ones who were in positions of authority. Had it not been for concerned Ordnance officers who were determined to provide a better tank for the troops and went over the heads of the obstructionists to have them overruled, the development which led to the Pershing and the shipment of the few which eventually got to combat zones probably would not have taken place.

Until now, no one has written this complete history. There are many reasons for this. As the author, Mr. R. P. Hunnicutt points out, it has taken ten years for him to piece it all together. He is to be commended for having had the patience to do so. Armor enthusiasts everywhere should welcome this most complete and authentic study of an interesting phase of World War II history.

Robert J. Icks
Colonel AUS — Retired

INTRODUCTION

Much has been written, since the end of World War II, on the amazing technical achievements in the development of German armored vehicles. Under the pressure of war, a wide variety of projects were initiated. A few were wildly impractical, but others resulted in designs which still remain as the basis for modern weapon development. Some of the new German tanks and self-propelled guns were introduced into battle when hardly past the prototype stage. Although they suffered numerous teething troubles, a profound impression was made upon the troops unfortunate enough to appose them. The publicity resulting from these actions, plus the availability of German development records after the cessation of hostilities, contributed to the intense interest in the subject. Even a quarter of a century later, new material is still being unearthed about the German wartime projects. In comparison, a much smaller amount of information has been published regarding the Allied development effort. This has been true, partly because of continuing security restrictions, and partly due to the difficulty in sorting the necessary material from the mountains of documents and test reports prepared during this period. Wholesale destruction of many such documents before declassification also contributed to the research problem.

Starting late compared to the Germans, the Allied research and development effort was no less intense and produced results no less remarkable. The objective of this book is to trace the course of one American development program, that of the T20 medium tank series, from the beginning to its conclusion in the postwar period.

The workhorse of American armored troops throughout the war was the M4 medium tank, nicknamed General Sherman. The Sherman, however, was a prewar design modified in light of the British battle experience. It proved to be extremely adaptable, serving throughout the war and long into the postwar era. Such long life resulted from its basic design flexibility, enabling it to absorb the many changes required to compete with newer vehicles. The source of many modifications used to modernize the Sherman was the new tank development program initiated early in 1942. With the Sherman in full production, attention was turned to the development of an improved tank using the latest components and design techniques. Funds were available to investigate a wide variety of weapons, power trains, and suspensions.

About this time, remarkable results were being obtained using an electric drive in the heavy tank T1E1 and a natural result was to adapt it for use in the new medium tank. An interesting parallel is noted here with the German development program where Dr. Ferdinand Porsche was using an electric drive in his version of the early Tiger and its descendent the Elefant self-propelled gun. Although the electric drive offered unique handling characteristics and excellent maneuverability, neither the American nor the German versions found favor with the combat troops. Lighter weight, less complex mechanical or hydraulic drives were preferred by the men who had to use them.

American tank doctrine required a highly mobile medium tank suitable for the exploitation role, but with sufficient firepower and protection to enable it also to operate in support of infantry. The T20 series sought to meet these objectives with considerable improvement over the Sherman. Unlike the prevailing Army Ground Force doctrine, the Armored Force and the Ordnance Department believed that the best antitank weapon was another tank. Following this line, the T20s were armed, first with the new 76 mm, and later the 90 mm high velocity guns. The latter offered an excellent balance of firepower, mobility, and protection in the T25E1 with their importance arranged in that order. Later efforts to match heavy German armor without increasing the engine power reduced the mobility of the T26 models. However, the troops in Europe, engaged by the high powered German tank and antitank guns, were calling for more and more armor protection. As a result, the T26E3 was standardized as the M26 (General Pershing) and even heavier armor protection was planned for the T26E5. With the standardization of the Pershing, there was no further interest in the T25E1 and the opportunity was lost to provide a successor to the Sherman with superior firepower and mobility.

The research for this volume spanned a ten year period involving travel to many parts of the United States and England. On many occasions when a particular subject seemed exhausted and nothing more was available, completely new material would turn up, in some cases reversing earlier conclusions. No doubt this will continue after publication. Any such new material or corrections of the present volume will be greatly appreciated by the author.

R. P. Hunnicutt
Belmont, California

March 1971

PART I

A NEW TANK FOR THE TROOPS

Pershings from Lieutenant Grimball's platoon of Company A, 14th Tank Battalion, 9th Armored Division wait for attack orders on March 1st, 1945 in a field outside Vettweiss, Germany.

ACTION AT REMAGEN

In the late morning of March 7th, 1945, the spearhead of the 9th Armored Division reached the heights overlooking Remagen on the western bank of the Rhine River. The troops from the 27th Armored Infantry and the 14th Tank Battalion had been pushing toward the Rhine ever since the offensive opened with the crossing of the Roer River on the morning of February 28th. During the first week of the advance, the leading tanks were the familiar Shermans widely used by all units in the U.S. Army. However, as the half-tracks of Lieutenant Timmerman's Company A, 27th Armored Infantry pulled out on the morning of the 7th, they were accompanied by a new silhouette, unlike any previously seen in American service. These tanks were wide and low and carried a massive turret mounting a long barreled cannon with a muzzle brake. This was the T26E3 (General Pershing), at that time classified as a heavy tank. The Pershing weighed approximately

45 tons and carried a 90 mm gun. The vehicle was designed to provide firepower and protection comparable to that of the German Tiger I.

Although a platoon of Pershings had been with Company A, 14th Tank Battalion since the crossing of the Roer River, little opportunity had arisen for their employment. Because of their greater weight and width, it was difficult for them to cross the narrow bridges spanning the numerous streams in this area. It was noted early in the drive that the wide tracks of the Pershings often damaged the prefabricated bridges installed by the Corps of Engineers. The necessary repairs often delayed the troops following behind. As a result, the Pershings were often forced to wait until other elements of the column had crossed over. By the morning of March 7th, most of these obstacles had been passed and for the final thrust toward Remagen, the platoon of Pershings was placed in front with Company A of the

9

Sergeant Key's Pershing knocked out by two heavy high explosive shells. The commander's vision cupola was blown off and can be seen on the ground beside the tank.

27th Armored Infantry. The five tank platoon, commanded by 1st Lieutenant John Grimball, of Columbia, South Carolina, had been reduced to four by this time.

The five T26E3s originally assigned to Grimball's platoon were serial numbers 22, 27, 28, 35 and 39. On the night of March 1st, Pershing number 22 was struck by two heavy high explosive shells while parked near a road junction east of the Roer River. The first shell, estimated as 15 cm in caliber, hit near the right rear sprocket damaging the running gear and starting a fire in the engine compartment. The crew, led by Platoon Sergeant Chester Key, dismounted to fight the fire when a second shell struck the turret just to the rear of the tank commander's hatch. Sergeant Key was instantly killed. The commander's cupola was blown off and landed 25 feet away. Later examination revealed that all but four of the turret mounting bolts were sheared. The vehicle was still being repaired on March 7th. An earlier casualty among Grimball's Pershings occurred when number 27 broke a piston shortly after crossing the Roer River. However, the engine was replaced and the tank returned to duty on March 5th, two days before the attack on Remagen.

When scouts from the leading infantry half-track reached the crest overlooking the Rhine Valley, they were astonished to see the Ludendorf railway bridge still intact, spanning the river on the other side of the town. Lieutenants Timmerman and Grimball came up to see for themselves and transmitted the information to the Task Force Commander, Lieutenant Colonel Engeman. When the discovery was reported to Brigadier General Hoge of Combat Command B, he ordered a rapid advance into the town by all troops and directed that a major effort be made to seize the bridge intact. Lieutenant Timmerman's infantry moved down the hill into Remagen with the tanks following along the road. Only minor opposition was encountered. A machine gun opened up on the infantry in the main square, but was quickly silenced by 90 mm fire from two of Lieutenant Grimball's tanks.

The push continued through the town until about 1400 hours when the tanks and infantry approached the western edge of the bridge. At this time, the German demolition team blew a large hole in the approach causeway blocking the entrance to the bridge for the tanks. On orders from Colonel Engeman, Lieutenant Grimball covered the bridge

10

The western end of the Ludendorf bridge on March 9th, two days after its capture. The hole blown in the approach causeway has been bridged and prisoners are being brought back across the Rhine. The eastern end of the bridge is shown below.

with machine gun fire to keep the Germans from using it and to interfere with their demolition procedures. As Timmerman and Lieutenant Mott of the Engineers were organizing their men for the crossing, the Germans set off an emergency demolition charge two-thirds of the way across the bridge. The huge structure lifted, timbers, dust and thick smoke mixed in the air, but it settled back intact although badly damaged.

Observing the movement of the infantry up to the bridge, the German machine gunners raked the approach from the two towers at the western end. Grimball's tank slammed a 90 mm high explosive shell into one of the towers and the machine gun fire slackened. The infantry began to move across the bridge cleaning out the towers as they went. About two-thirds of the way across, they were taken under fire by snipers from a half submerged barge about 200 yards up stream. A Sherman tank, which had now joined the Pershings, blasted the barge with its 75 mm gun. As the engineers cut away the demolition charges, the infantry completed their crossing. Un-aware of the situation, a German train steamed into

11

The tanks from Lieutenant Grimball's platoon being loaded on ferries constructed from bridge pontons on March 12, 1945.

The bridge area was still under enemy artillery fire as the ferries were pushed across the Rhine.

view on the far side of the river and was enthusiastically destroyed by the tanks, exploding the boiler and scattering the troops which poured out of the cars.

The bridgehead on the eastern side of the river was secured by the infantry and feverish efforts followed to fill the hole in the approach causeway to permit supporting tanks to cross over. This work was completed by nightfall and during the night two platoons of Shermans and some tank destroyers crossed to reinforce the infantry. Because of their wider dimensions and heavier weight, the Pershings were not permitted to cross the damaged bridge. Five days later, on March 12th, barges were improvised by joining five bridge pontons and the Pershings were ferried to the eastern shore.

The 3rd Armored Division first introduced the Pershing into action with the attack across the Roer River ten days before Remagen. However, the seizure of the vital bridge on the Rhine was of greater significance as far as the course of the war was concerned. Combat Command B of the 9th Armored Division was also the first to use the Pershing in a full platoon. The other organizations equipped from the original shipment distributed their tanks in ones and twos to separate companies, but in the 14th Tank Battalion all five Pershings were concentrated in one platoon under Lieutenant Grimball.

12

T26E3, serial number 55, shown here is representative of the early production Pershings. This particular tank was under test at Aberdeen Proving Ground.

THE ZEBRA MISSION

The full story of the Pershing's introduction to combat must be traced back to the United States in the fall of 1944. Production of the T26E3 was just starting at the Fisher Tank Arsenal and Major General Barnes, Chief of Ordnance Department Research and Development, was convinced that these vehicles were urgently needed in the European Theater of Operations. He proposed that of the first 40 vehicles produced, 20 should be sent directly to Europe and simultaneously 20 to Fort Knox for tests by the Armored Board. Army Ground Forces objected to this procedure, saying that the shipment to Europe should be delayed until the tanks were tested and approved by the Armored Board. Convinced that time was too short for such delays, General Barnes threatened to carry his argument to General Marshall, if necessary. Major General Russell Maxwell, Assistant Chief of Staff, G4, agreed with his argument and 20 T26E3s were shipped to Antwerp, Belgium in January 1945.

To assist in the rapid introduction of the Pershing as well as several other new weapons, General Barnes led a technical mission to Europe. This mission, code name Zebra, arrived in Paris on February 9th, 1945. Besides General Barnes, the group included Colonel Joseph Colby and Captain Elmer Gray of the Tank Automotive Command, Colonel George Dean of the Army Ground Forces New Developments Division, Captain Gifford Griffin from Aberdeen Proving Ground and two civilians. The latter were W. A. (Bill) Shaw, a representative from the Fisher Tank Arsenal, who had been with the T26E3 since its original conception, and L. R. (Slim) Price from Aberdeen Proving Ground, an expert on the 90 mm gun mounted in the Pershing. After a meeting with General Eisenhower, it was decided to get the 20 tanks into action as soon as possible and they were all assigned to the 12th Army group where General Bradley sent them to First Army, dividing them equally between the 3rd and the 9th Armored Divisions.

A Pershing from the first shipment being loaded aboard a modified M25 tank transporter at Antwerp, Belgium, 9 February 1945.

After conferences with General Barnes and Colonel Colby, Captain Gray was told that he had the ball and to carry it. With these instructions, Gray and his team headed for Antwerp on February 11, 1945. Captain Griffin had been assigned to accompany the shipment of tanks by sea with definite orders that all boxes and equipment were to be unloaded at the port, but nothing was to be turned over to the Transportation Corps for movement. When Gray arrived at Antwerp, his worst fears were realized. Captain Griffin had been outranked and all of the boxed goods had been loaded into 15 freight cars with the promise that they would travel as one train. However, in line with previous experience, the cars were quickly scattered, requiring the assignment of a Transportation officer to track them down. After two weeks of hard work, this officer managed to round up all 15 cars and escort them to the parts depot.

Wreckage of a German V-1 in "buzz bomb alley".

Antwerp was being heavily bombarded by the German V-1 "buzz bombs" during this period. To avoid possible loss, the tanks and equipment were loaded on modified M25 tank transporters and moved to Brussels. These transporters had the trailer reinforced to handle the Pershing's 45 tons and ramps were added over the rear tires to permit loading the wider tanks. In Brussels, Captain Gray received word to meet General Barnes and Colonel Colby at Eagle Headquarters in Namur, France. This was the 12th Army Group Headquarters of General Omar Bradley. Gray headed for the rendezvous, but had to continue on to First Army Headquarters at Spa where Barnes and Colby were conferring with General Hodges. When he arrived, Gray received orders from Colonel Colby to move the tanks immediately to Aachen, Germany, where the 559th Heavy Maintenance Tank Company would prepare them for combat. Returning to Brussels, the transporters were checked and prepared for the long road trip. One of Captain Gray's less orthodox preparations involved removing the engine mufflers thus insuring maximum power for a fast passage. A day was consumed with the preparations and obtaining road clearances for the convoy through the British sector. Some confusion was associated with the latter and the Military Police attempted to stop the convoy when it moved out on Saturday morning, February 17th. The details of the confrontation are somewhat clouded, but rumor has it that the M.P.s, led by a major, were threatened with being run over by the loaded tank transporters. Whatever the truth of the matter, the convoy arrived at its destination at 1700 hours the same day, as Gray states in his report, "without mishap or misgivings." It had been raining continuously and was bitterly cold, but in spite of the weather, members of the 559th started to prepare the tanks for combat.

General Barnes and Colonel Colby had made a wise selection in the choice of Captain Gray for this project. The success of the rapid movement from Antwerp to Aachen and of many operations to follow were due in great measure to his efforts. His skill in cutting red tape and getting a job done had been proven during an earlier mission introducing heavy artillery pieces in the summer of 1944. Although Elmer Gray insists that he went through channels, by his own admission these channels were considerably wider and deeper following his passage.

At Aachen, the crews assigned to ten of the tanks had arrived from the 3rd Armored Division and classes were formed for their instruction. Captain Gray instructed the group on elementary mechanics and the components of the vehicle, while Captain Griffin covered operating procedures. Slim Price held a class on gunnery for the tank commanders, gunners and loaders.

The tanks were checked out and the crews trained despite the cold, wet weather at Aachen. The low temperature is apparent from the appearance (above) of the General Motors representative, Bill Shaw, taken during the training program.

The tall officer in the photograph below is Lieutenant John Grimball who led the Pershings at Remagen.

Gunnery training at Aachen introduced the veteran crews to the 90 mm gun. The flash and smoke were much greater with the 90 than with the smaller tank guns.

By February 20th the crews from the 3rd Armored Division had completed their schooling and their tanks had been checked out mechanically. However, as yet the guns had not been boresighted and test fired. Slim Price selected a firing range and supervised the crews in this work. Each tank crew fired 28 rounds to thoroughly familiarize themselves with the weapon. In the meantime, crews had been moving in from the 9th Armored Division to receive the remaining ten tanks. These men underwent a similar training program while their tanks were being brought up to combat readiness. In addition, schools were held for the maintenance battalions in both the 3rd and 9th Armored Divisions.

Through all of his instruction on the 90, Slim Price emphasized the importance of properly boresighting the weapon. The gun was extremely accurate, but this advantage could easily be lost by careless boresighting or failure to minimize the effect of backlash in the traverse and elevation gears. Price checked the original 20 tanks after boresighting by their crews and found only one met his standard.

Even with the muzzle brake acting as a deflector, smoke and dust created problems by obscuring the target during firing of the 90. Note the empty shell case on the ground from the previously fired round.

ZEBRA MISSION

Assignment of 20 T26E3 Tanks

3rd Armored Division

32nd Armored Regiment				33rd Armored Regiment		
Serial No.	Registration No.	Company		Serial No.	Registration No.	Company
26	30119836	E		24	30119834	D
31	30119841	H		25	30119835	H
33	30119843	G		37	30119847	I
34	30119844	I		38	30119848	F
36	30119846	D		40	30119850	E

9th Armored Division

14th Tank Battalion				19th Tank Battalion		
Serial No.	Registration No.	Company		Serial No.	Registration No.	Company
22	30119832	A		23	30119833	C
27	30119837	A		29	30119839	B
28	30119838	A		30	30119840	A
35	30119845	A		32	30119842	B
39	30119849	A		41	30119851	C

The 3rd Armored Division retained the 1942 organizational structure with two armored regiments while the 9th Armored Division followed the 1943 tables of organization replacing the two regiments with three tank battalions.

After correcting the other 19 tanks, he demonstrated the accurate shooting possible with the 90. His method, which he passed on to his students was not to just aim at the enemy tank, but to aim at a particular spot on that tank and his demonstration showed it was possible to hit that spot. Later in the mission he impressed some of Patton's tank crews by using German helmets as targets and picking them off with single shots from the 90 at a range of 625 yards. Such performance quickly overcame any objections that veteran tankers might have had to receiving gunnery instruction from a civilian. In fact, the veterans rapidly duplicated his markmanship once they got the "feel" of the weapon.

During the initial test firing, a malfunction occurred, typical of those which plague equipment still under development. When firing the first round from a cold gun, the recoil was insufficient to engage the cam designed to open the breech and eject the empty shell case. As the gun warmed up, the length of recoil increased and the breechblock operated properly. However, such a failure on the first shot could be a matter of life or death for the tank crew if that shot missed or more than one enemy tank was present. The time required to manually open the breech and eject the shell case would give the enemy ample time to return the fire. The problem arose from the fact that the cam was designed for the 90 mm gun as mounted in the T26E1. This gun was not fitted with a muzzle brake. The addition of the muzzle brake on the T26E3 reduced the recoil to the point where it would not trip the cam if the gun was cold. Gray and Price solved the problem by removing ½ inch from the cam without changing its contour. The breech block would then operate under any firing conditions.

Boresighting and test firing were completed on February 23rd and the tanks were ready for combat.

On February 25th the 3rd Armored Division was committed to battle, attacking across the Roer River. Three days later the 9th followed. During this period the Ordnance team made their home with the 3rd Armored Division's maintenance battalion in Düren, Germany and from there followed the tanks into action, supervised necessary repairs, and recorded their performance.

Captain Gray drove from Düren to Elsdorf on the 28th to check on a 3rd Armored T26E3 reportedly knocked out two days before. This tank, the first lost in action, proved to be serial number 38 assigned to F company, 33rd Armored Regiment. The Pershing was named "Fireball" with the name painted

The first loss. T26E3 number 38, "Fireball", knocked out by a Tiger at Elsdorf. The 8.8 cm hole through the machine gun port can be seen above and in closeup below. The round objects on the front armor are the belly escape hatch covers.

on both sides of the vehicle. The tank had been positioned behind a roadblock to watch for enemy movement. This turned out to be a poor location with the fires burning in the vicinity. In the darkness, flames from a burning coal pile silhouetted the turret which was exposed above the road block. A German Tiger tank, concealed behind the corner of a building, fired three times at the turret at a range of about 100 yards. The first 8.8 cm shot penetrated through the coaxial machine gun port, spun around inside the turret killing the gunner and the loader. The second hit the muzzle brake and the end of the gun tube jarring off the round that was in the chamber. The discharge of this shell caused the barrel to swell at about the halfway point even though the projectile went on out the tube. A third shot glanced off the upper right-hand side of the turret tearing away the cupola hatch cover which had been left open. The Tiger then backed up, immobilizing itself on a pile of debri and was abandoned. The loader of the Tiger was later captured and confirmed that his tank had done the firing.

The Ordnance team surveys the damaged "Fireball". The bulge in the 90 mm gun tube is visible where the paint has spalled off and the remains of the smashed coaxial machine gun are on the right fender. The scuff mark where an 8.8 cm projectile bounced off the turret can be seen above and in closeup at the left. The damaged muzzle brake is at the left below and the left side name marking is shown below.

FIREBALL

"Fireball" repaired and ready to return to action on March 7th. Note that the coaxial machine gun port has been repaired by welding and a new gun tube is installed.

"Fireball" was promptly avenged the following day when Pershing number 40 assigned to Company E 33rd Armored Regiment, knocked out and burned a Tiger I and two Panzer IVs, also at Elsdorf. Four shots were fired and registered on the Tiger at a range of approximately 900 yards. The first, a T30E16 HVAP projectile, destroyed the final drive. The second, a T33 shot, hit the bottom of the gun mantlet next to the hull penetrating the turret causing an explosion. Two other hits by high explosive were ineffective. The two Panzer IVs were knocked out and burned at a range of 1200 yards with one round each of T33, but an additional two rounds of high explosive were used to destroy the crews as they

escaped. Another Panzer IV was later destroyed by this crew during the drive to Cologne.

The damaged "Fireball" was immediately evacuated to the division maintenance battalion at Düren for repairs. A truck was dispatched to the 310th Ordnance Battalion which was handling the spare parts brought along on the mission. Unfortunately, these parts did not include 90 mm guns and "Fireball's" cannon was damaged beyond all hope. A phone call to Paris saved the situation by obtaining a spare 90 mm gun tube from an M36 tank destroyer. Repairs were then completed with the shot holes welded up and the new parts installed. "Fireball" was recommitted to action on March 7th.

Fireball's avenger, Pershing number 40, which destroyed a Tiger I and two Panzer IVs also at Elsdorf. The tank commander is second from the left.

20

Slim Price, at the left, and the Pershing tank commander stand in front of the Tiger knocked out at Elsdorf. The 90 mm hole through the lower gun shield can be seen just to the left of the tank commander's face. A closeup of this hit is shown below. Below at the right is a view of the hit on the final drive by a 90 mm T30E16 HVAP shot.

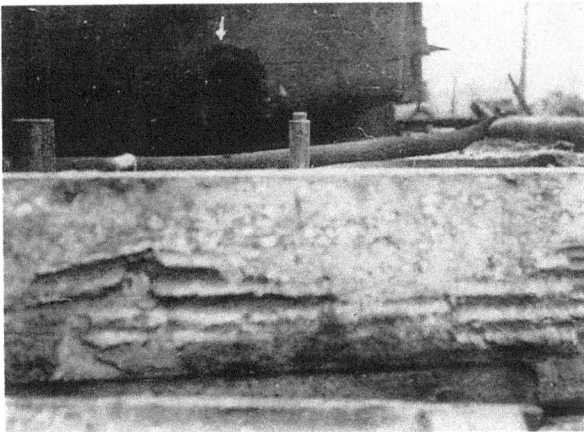

The two Panzer IVs destroyed in the action at Elsdorf are shown below. Both were knocked out with single rounds of T33 ammunition, but considerable damage was done to the running gear and external stowage by high explosive.

By this time action had started to pop on all sides. As related earlier, Sergeant Key's tank (serial number 22) was disabled on March 1st by two rounds of high explosive. They caused extensive damage to the turret and running gear. The blast from the first shell destroyed the stowage boxes, sand shields, and many of the end connectors, on the new T80E1 double pin tracks. This was one of the first tanks equipped with the new track. The right rear road wheels were damaged and the sprocket hub split. When the second shell blew off the commander's cupola, it also destroyed the fire control instruments, the intercommunication system, and damaged the turret race. Several weeks were required to restore the tank to combat readiness.

The Ordnance team also had its hands full with a number of non-combat casualties. Two tanks suffered main engine failures. The first was serial number 27 from Lieutenant Grimball's platoon followed by number 37 assigned to Company I of the 33rd Armored Regiment. Both vehicles were repaired and recommitted to action on March 5th. Another Pershing threw a connecting rod through the crankcase of its auxiliary engine and still another was hanging half on and half off a bridge over the Erft Canal. All of these were recovered and repaired.

On March 6th Captain Gray was with the 3rd Armored in Cologne searching for a T26E3 that was reported knocked out, but was unable to find it due to bitter fighting taking place in the area. The following day Cologne was captured and he finally located the tank at the little suburb town of Niehl, north of Cologne on the banks of the Rhine. This was Pershing number 25 belonging to Company H of the 33rd Armored Regiment and it had been knocked out by a self-propelled 8.8 cm gun (Nashorn) at under 300 yards range. The projectile penetrated the lower front armor, passed between the driver's legs and set the turret on fire. The crew successfully abandoned the vehicle before the ammunition blew up gutting the turret. Although the tank was repairable, it would have taken several months to make it combat serviceable. Considering the severe shortage of spare parts, it was evacuated to the rear for cannibalization. Out of the original 20 tanks of the Zebra Mission, this was the only one which did not finish the war on active service.

Sergeant Key's tank, Pershing number 22, is shown in this photo sequence. The new type T80E1 double pin track is clearly revealed in the two middle pictures. The lower photograph shows the damaged end connectors and drive sprocket. All of this damage resulted from the first hit.

The second hit on number 22 caused extensive damage inside and outside the turret in addition to blowing off the vision cupola. Looking into the turret at the right above, the azimuth indicator and other fire control equipment is seen to be damaged from the blast.

Engine replacement and repairs were carried out under field conditions with crude improvised equipment. Despite this, a remarkable record was attained in keeping the tanks serviceable.

Pershing number 25 knocked out by a self-propelled 8.8 cm gun near Cologne. The hole through the lower front armor on the driver's side can be seen at the bottom of the photo at the right, a closeup of the same shot hole is shown at the right below. The turret was completely burned out, but the remainder of the vehicle was undamaged.

In the fighting for Cologne on March 6th, Pershing number 26 commanded by Sergeant Early of Company E, 32nd Armored Regiment, knocked out and burned a Panther in front of the Cologne Cathedral. One round of T33 ammunition struck the base of the turret just to the left of the gun tube. Another projectile went through the right sponson and a third penetrated the right front side of the hull. During this same drive T26E3 number 36, from Company D, 32nd Armored Regiment, knocked out a Tiger I using two rounds of T33 ammunition. From the same regiment, Company G's Pershing (serial number 33) killed a Panzer IV at Manheim with three rounds of M82 through the side.

Some of the most dangerous resistance around Cologne and other approaches to the Rhine came from the heavy antiaircraft defences. These weapons were deadly when used as antitank guns and the Pershing's 90 mm high explosive shells proved highly effective in dealing with them. The Germans also improvised fixed antitank mounts using the guns from many of their fighting vehicles immobilized by the fuel shortage.

Captains Gray and Griffin were at Remagen on March 12th observing the preparations for ferrying Lieutenant Grimball's Pershings across the Rhine. Groups of five bridge pontons were lashed together to form barges and a floor installed. The tanks were then driven on and pushed across the river using large outboard motors. Griffin and Gray crossed the Rhine on the 13th and confirmed that the vehicles were in good condition. A week later Captain Gray was informed that additional shipments of T36E3 tanks were expected, so he and his group returned to Paris.

Above: Captain Elmer Gray, at right, and Bill Shaw, the General Motors representative, with the 3rd Armored Division in Cologne.
Below: The Panther destroyed by Pershing number 26 still burning in front of the Cologne Cathedral.

Above: The Tiger I which knocked out "Fireball" (Pershing number 38), immobilized on a pile of debris and abandoned near Elsdorf.

Below: Pershing from C Company, 19th Tank Battalion, 9th Armored Division moves along the road between Thum and Ginnick, Germany, March 1st, 1945. Note the mine exploder in the background.

The 8.8 cm Pak 43 from a Jagdpanther emplaced as a fixed antitank gun. The weapon was mounted on a concrete base with 360 degree traverse. The numbers used in the "clock" system of fire control can be seen around the edge of the gun pit. This gun was destroyed by high explosive shells from Pershings of the 9th Armored Division. The covered ammunition storage for the 8.8 cm rounds is shown below.

A few days before the return to Paris, the first so-called "Super Pershing" arrived from the United States. It was the result of an effort to improve the Pershing's firepower by installing the new T15E1 90 mm gun in T26E1, number 1. This weapon had a much higher muzzle velocity than the standard 90 mm gun M3 and was comparable in performance to the German 8.8 cm KwK 43 mounted in the King Tiger. With its new armament, the tank was redesignated as the T26E4, temporary pilot number 1, but retained its old registration number, 30103292. This vehicle is described in more detail later in the Development History.

After proof firing at Aberdeen, the tank was shipped to Europe and delivered to the 3rd Armored Division's maintenance battalion on March 15, 1945. Before leaving Aberdeen, Slim Price had installed a special telescopic sight designed for use with the high velocity weapon. On arrival, however, the new telescope was missing and in its place was the standard M71C sight normally used with the 90 mm gun M3. Captain Gray checked back through all of the units handling the tank since its arrival in Europe in an effort to find the missing sight. He finally flew to Paris and located a Lieutenant McDougal who had been in charge of the tank before it was shipped to the 3rd Armored. McDougal informed him that the M71C scope was installed when the vehicle was received from the United States. Obviously, an overzealous depot crew had installed a new M71C sight when preparing the tank for shipment. Since it proved impossible to obtain a new special sight, Price had to work out a range data sheet for use with the standard scope.

About a week earlier, Gray had rescued the special ammunition for the T15E1 gun. These fixed rounds, approximately 50 inches long, had been mistakenly shipped to the 635th Tank Destroyer Battalion. The 635th was evaluating the new 90 mm T8 antitank gun on the towed carriage T5E2. This was another new weapon brought over by the Zebra Mission and it used the same standard ammunition as the M3 gun. The misplaced rounds came to Gray's attention when the 635th telephoned to ask why the ammunition was 12½ inches too long for their gun.

In the meantime the maintenance battalion, with Price's help, had been preparing the tank for combat. Since it had such a powerful gun, they decided to improve the armor protection to put it on more equal terms with the King Tiger. A piece of armor was flame cut from the 80 mm glacis plate of a captured Panther and welded to the front of the gun shield. The front hull armor was reinforced with a double layer of plate greatly increasing its slope. Some additional stowage boxes were installed and the "Super Pershing" was ready to go about a week after it was received. With a crew assigned from the 3rd Armored, the tank was moved forward in the hope of slugging it out with a King Tiger, but the war ended before such an encounter occurred.

The "Super Pershing" as fitted up by the 3rd Armored Division maintenance battalion. The two cylinders over the gun mount contain the coil springs necessary to compensate for the weight of the long heavy gun barrel.

Details of the armor plate added to the "Super Pershing" are shown in these photographs taken during the modification. The 80 mm plate taken from a Panther can be seen above welded to the gun shield. The other views show the double plate added to the front hull. Note the additional reinforcement above the bow machine gun mount. The effect of the additional weight added to the front of the tank can be noted above and in the photograph on the previous page. The extra armor plus the longer heavier gun barrel badly overloaded the front torsion bar springs causing the tank to tilt forward.

The end of the line for the "Super Pershing". The photographs on this and the following page show the final resting place of the tank in a vehicle dump near Kassel, Germany. These pictures were taken by Colonel G. B. Jarrett in June 1945.

Note that additional armor plate has been attached to each side of the gun shield since the tank was originally modified. These wing like shields were intended to increase the protection of the upper front sides of the turret. Extra armor was also added over the equilibrator spring cylinders. The double door revolving loader's hatch, characteristic of the T26E1, is clearly visible.

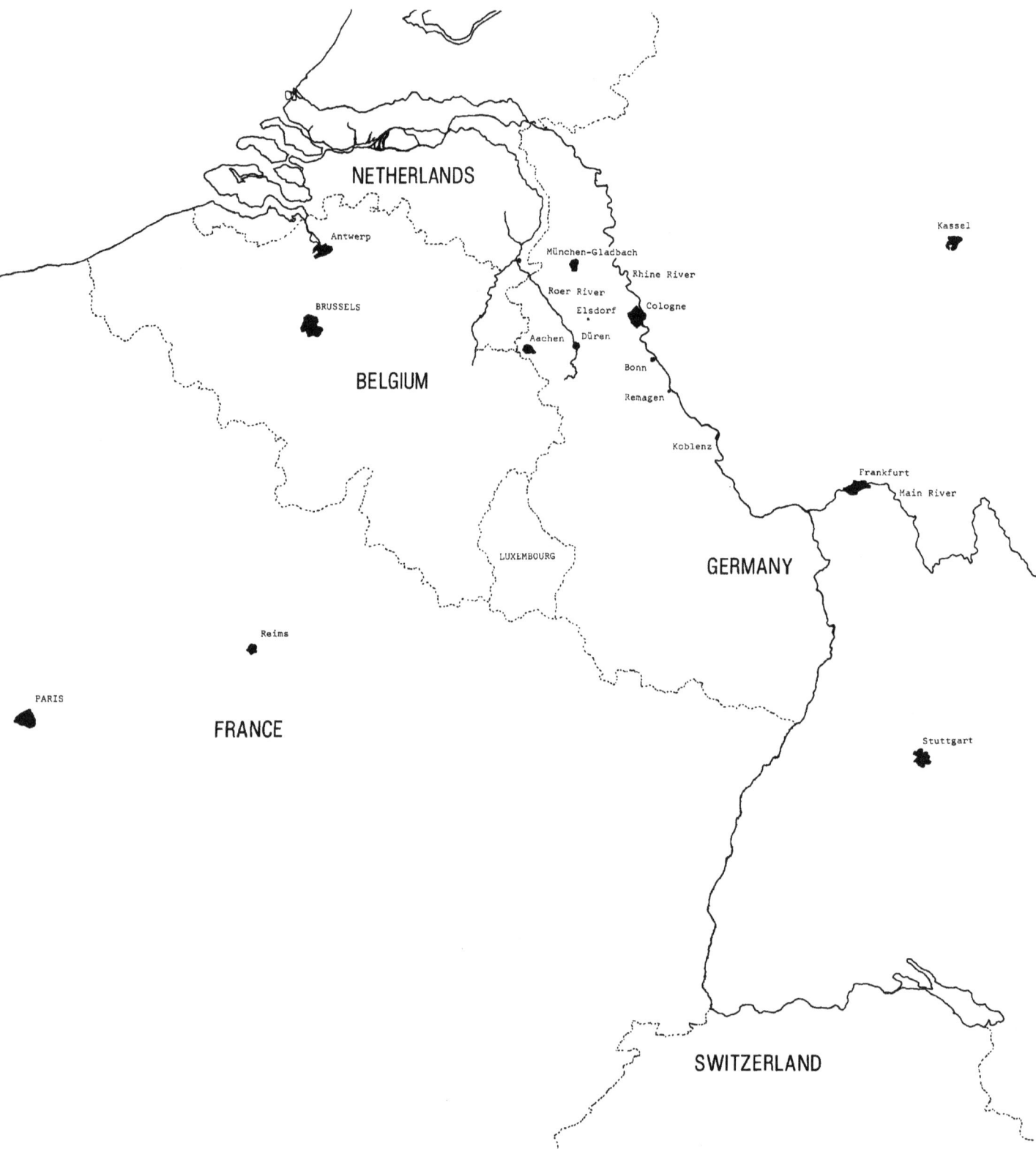

Zebra Mission, Area of Operations

Early production Medium Tank M26 (T26E3)

March 25th found the team split up between Antwerp and Le Havre to supervise the unloading and transporting of a new shipment of tanks to their allotted destinations. Shortly afterward, Bill Shaw became extremely ill at Le Havre and was rushed to a hospital in Paris. Captain Gray, complying with doctor's orders, had him flown back to the United States.

The Ninth Army was allocated the first group of 40 Pershings (24 from Antwerp plus 16 from Le Havre) from the new shipment. They were divided with 22 tanks going to the 2nd Armored Division and 18 to the 5th Armored Division. Captain Griffin and Slim Price were sent ahead to the 2nd Armored Division to arrange training for the tank crews. Captain Gray spent several days setting up the distribution of spare parts and trying to expedite the modifications of additional M25 tank transporters. He visited Ninth Army Headquarters on April 2nd and picked up information on the approximate location of the 2nd Armored Division. Reaching the division after a wild ride through 175 miles of very loosely held territory, Gray rejoined Griffin and Price. Very little time had been available for the latter to train the tank crews. After Price boresighted the guns, 45 minutes were allotted for briefing the men. They then went into action without firing a single practice shot.

The Pershings for the 5th Armored Division were now arriving at München Gladbach, so the team returned there to set up their school. Classes were held for the division maintenance battalion as well as the members of the 536th Heavy Maintenance Tank Company who were processing the tanks.

Unloading the second shipment of Pershings at Antwerp in March 1945. The modern facilities available for spare parts storage and distribution are shown below.

Pershing convoy enroute to the 2nd Armored Division approaches the Rhine near Wesel, Germany on March 30, 1945. Below, the tanks are crossing the Rhine and refueling on the far side. The Technician 4th Grade in the right photograph wears the insignia of the 2nd Armored Division.

T26E4, Temporary Pilot No. 1

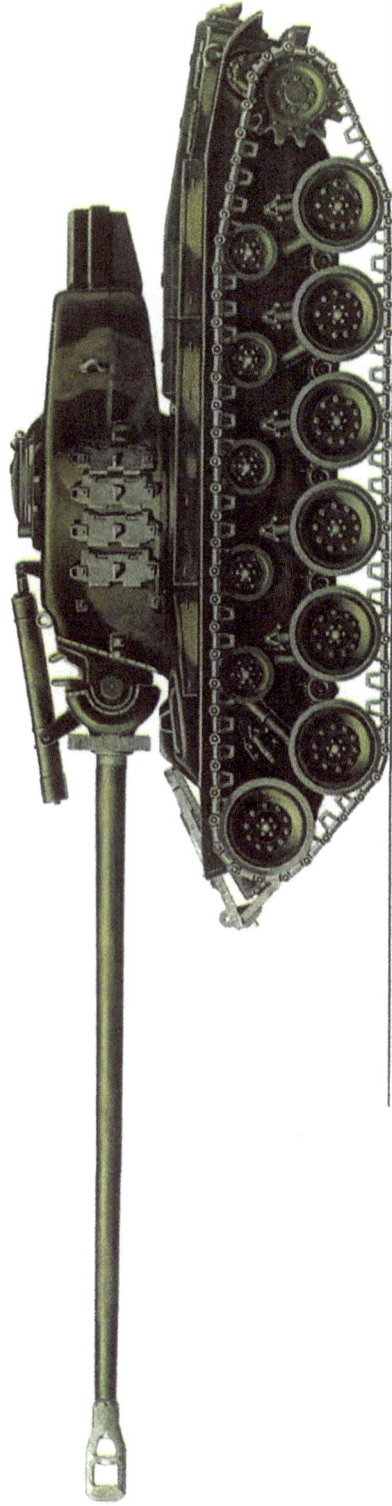

The "Super Pershing" after fitting with extra armor in the European Theater

Scale 1:48

36

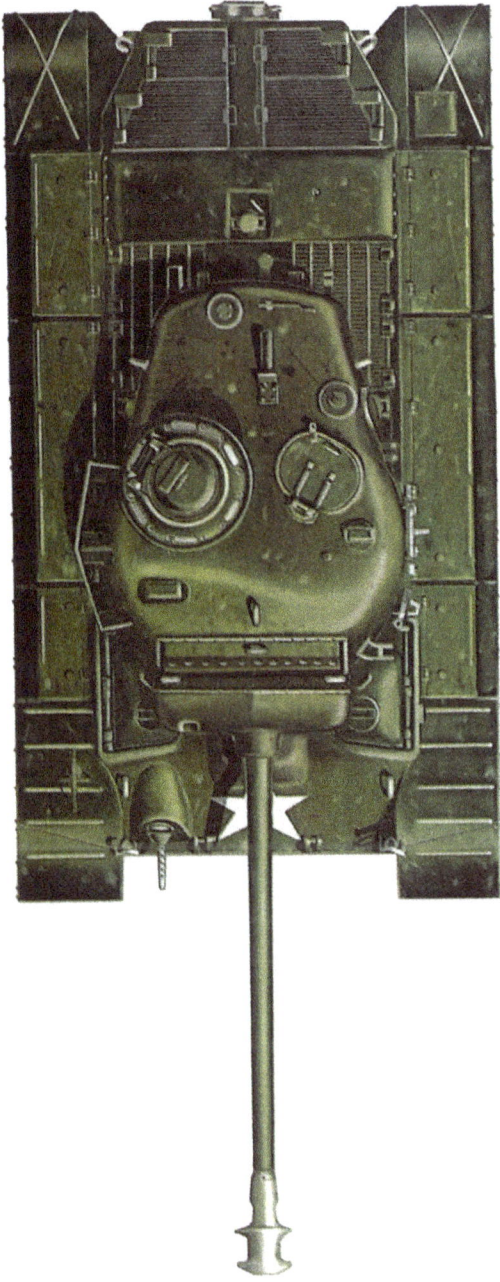

Early production Medium Tank M26 (T26E3)

37

Pershings for Patton, tanks assigned to the 11th Armored Division arrive at Third Army near Frankfurt am Main. Chalk marks on the front armor indicate the oil was changed on April 14, 1945.

Thirty T26E3s from the new shipment were allocated to the 11th Armored Division in Patton's Third Army. Gray and Price left for Third Army Headquarters on April 4th to make the usual arrangements to receive and issue the tanks. Captain Griffin moved forward with the 5th Armored tanks to brief their crews. When Gray and Price checked in at Third Army Headquarters in Frankfurt, they discovered that the tanks had not yet arrived. Phone calls to Paris to speed up the shipment revealed that the Corps of Engineers could not decide on a suitable bridge to cross the Rhine. After considerable "buck passing", permission was obtained to use a bridge in the Seventh Army area. However, by this time the evacuation company handling the movement had already found the bridge and crossed with 15 tanks on

April 12. Permission to use the bridge came through the following day.

Classes were set up for the 11th Armored tank crews as well as for members of the maintenance battalion and several Ordnance companies. Three hundred rounds of ammunition were drawn and the crews were allowed to fire ten rounds per tank. It proved impossible to locate a suitable 1000 yard range in the area, so Price had to settle for firing across a small lake for a distance of 625 yards. This was where he demonstrated his markmanship with the 90 using the German helmets as targets.

The crews from the 11th Armored completed their training and were committed to battle on April 21. By this time, additional Ordnance teams had arrived from the U.S.A. and further training was turned over to them. The original team returned to Paris and then to the United States, arriving in New York on April 26, 1945. The flow of Pershings to Europe continued until by VE Day there were 310 in the Theater of which 200 had been issued to the troops.

Captain Gray on the autobahn with his team enroute to Frankfurt.

Demonstration firing for the Third Army across a small lake near Frankfurt. The smoke produced by firing the 90 mm gun is apparent in these photographs. The effect of 90 mm high explosive shells in the target area is shown at the right below.

Medium Tank M45 (T26E2)

Medium Tank M26A1

Scale 1:48

40

The problem: A Sherman tank destroyed by the Japanese at Okinawa. This tank had its turret blown off by a satchel charge after being knocked out and abandoned.

PERSHINGS TO OKINAWA

The war in Europe was drawing to a close and greater attention could be given to the situation in the Pacific. Fierce fighting was in progress on Okinawa with U.S. tank losses far exceeding expectations. In particular, the Japanese were using their small high velocity 47 mm antitank guns with deadly effect. The small size of the weapon made it easy to conceal and its armor piercing projectile could penetrate any part of a Sherman except the glacis plate. At Kakazu, on the morning of April 19th, Company A of the 193rd Tank Battalion lost four Shermans to a single 47 mm gun. By the end of May, tank losses for the four Army tank battalions reached 221, of which 94 were damaged beyond repair. In the face of this heavy opposition, the tankers were demanding greater firepower and more armor protection. As an interim measure, the Ordnance maintenance companies were welding track links and extra sections of armor to the sides of the Shermans. However, the best answer appeared to be the Pershing, now standardized as the M26.

A closeup view of the Sherman shows the extra armor plate added to the sides in an effort to gain protection from the 47 mm Japanese antitank gun.

Less than three weeks after his return, Captain Gray was called to a meeting in the office of Brigadier General Borden, Chief of the New Developments Division, Washington, D.C. He was advised that a mission was being formed for the immediate shipment of 12 Pershing tanks with tools and ammunition to Naha, Okinawa.

Because of his previous experience introducing the tank in Germany, Gray was asked to head the mission and institute a training program in the use and care of the vehicle for both the operating personnel and Ordnance maintenance crews. Changes or modifications necessary for use in the Pacific Theater

were also to be noted. Returning to Detroit on May 16th, Gray was joined by Captain William Tauss, Engineer and Project Officer for the M26 tanks, who was also assigned to the mission. With preparations complete, the two captains departed for Seattle, arriving on May 24th. Having learned from earlier missions to take nothing for granted, they began to check the various shipments which were to accompany the tanks to their destination. This check revealed that many parts and major replacement units were back ordered. Previous experience had shown that it was highly unlikely that any such back ordered items would arrive in time to be of use to the mission. Fortunately, a quick search discovered two M26s at the Richmond Tank Depot and a telegram to Washington obtained permission to cannibalize these vehicles. Two complete power trains, one auxiliary engine and generator set, and one radiator assembly were removed and trucked to Mukiltea, Washington, where the ship was being loaded with ammunition.

The ship, the S.S. Katherine D. Sherwood, sailed on May 31st and since their orders read, "Direct shipment", they happily believed they would be in Okinawa not later than June 30th. Unfortunately, they did not reckon with the strained communications in the Pacific Theater. This became apparent at the first stop at Eniwetok. Expecting a one or two day delay, they were informed that the ship would leave in ten to fifteen days and the destination would then be Ulithi.

M26 tanks being lifted from the hold of the S.S. Katherine D. Sherwood off Naha, Okinawa. Lack of docking facilities required that the tanks be transferred to landing craft in the open sea.

Pershings being lowered into the Mark VI LCTs off Naha. The tanks were fueled and the engines started to enable them to maneuver under their own power.

When Gray tried to explain their mission to the port director and inform him that the tanks were being shipped on a high priority basis to alleviate the situation on Okinawa, he replied that since the ship had a low priority there was not much he could do. A radio message was sent to General Barnes' office and a letter to Colonel Colby explaining the situation.

After sweating it out for ten days, the ship hoisted anchor on June 27th and headed for Ulithi, 1350 miles away. It was the same story all over again, with an expected delay of ten days to two weeks. They were further informed that ships were sent to Okinawa only as they were called forward by name from that area. A few days were spent trying to get the ship routed to Okinawa and the two captains sent an informative letter through channels to General Stillwell. After 12 days at Ulithi they left in convoy on July 15th and arriving Okinawa on July 21. By this time all action had ceased and the build up was starting in preparation for the invasion of Japan.

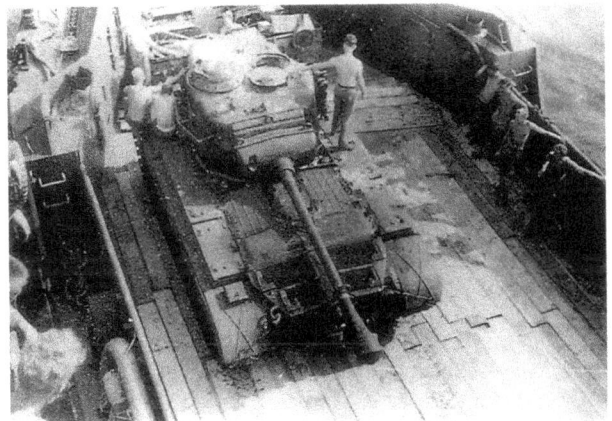

The first two Pershings ashore at Naha on July 31, 1945.

Unlike Europe, the unloading facilities at Okinawa were extremely crude, requiring the tanks to be loaded into LCTs and landed on the open beach. Further delays occurred when it was discovered that only the later mark LCTs could handle the large Pershings. Finally, late on July 30th, the first two tanks were transferred to the LCTs and landed on the beach at Naha the following day. Two more tanks were moved ashore before all operations were stopped by typhoon warnings from the local weather station. The remaining vehicles were lashed down and the ship put out to sea to ride out the storm. They returned to Naha on August 4th and the remaining eight tanks were unloaded.

Arrangements were made to train personnel from the 193rd and 711th Tank Battalions with six Pershings to go to each unit. The 81st and the 293rd Heavy Maintenance Tank Companies were to be included in the training program.

On August 10th, the first class was conducted by Gray, now promoted to Major, at the 193rd Tank Battalion. That night, word was received that Japan might accept the terms laid down at the Potsdam Conference and orders were issued to discontinue all training and stand by. Tenth Army had instructions to be ready for overseas movement to Korea in 17 days. V-J Day on August 15th closed the chapter. Since the tanks would no longer be used in combat, the mission was considered completed and Gray and Tauss were free to return to the U.S.A. They left Okinawa August 21, arriving at San Pedro, California, Sept. 14, 1945. The war was finished and, except for a period in Korea, so was the combat history of the Pershing tank.

One of the first four Pershings landed at Okinawa on July 31, 1945.

Display of the Pershing with other new equipment on Okinawa. The officer in the righthand photograph is Captain William Tauss, Project Officer for the M26.

Although poor in quality, the two photographs here are included because of their rarity. To the author's knowledge, they are the only photographs available of the Pershing test firing at Okinawa.

MISSION TO OKINAWA

Serial and Registration Numbers of 12 M26 Tanks

Serial No.	Registration No.	Serial No.	Registration No.
586	30127332	617	30127363
603	30127349	621	30127367
606	30127352	623	30127369
608	30127354	625	30127371
609	30127355	629	30127375
614	30127360	632	30127378

All of these tanks were manufactured by the Fisher Body Division of General Motors Corporation.

PART II

THE DEVELOPMENT HISTORY

At the left is an early artist's conception of the T20 design. Note the resemblance to the gun shield of the 3 inch gun motor carriage M10 on which the T80 gun mount was based. The mock-up of the T20 as presented in May 1942 is shown at right with the T79 gun mount.

REQUIREMENT FOR A NEW TANK

The development program which finally evolved the Pershing began in the spring of 1942. When Colonel Joseph Colby returned from Africa in April, the medium tank M4 (General Sherman) was just coming into full production. At that time it was considered superior to any tank on the African battlefield. However, German tank development showed a trend toward greater firepower and improved armor protection. This was illustrated by appearance of the Panzer III J with strengthened front armor and a long 5 cm L/60 cannon. Nineteen of these tanks were on strength with the Afrika Korps for the battle of Gazala in late May 1942. Shortly thereafter, the first

Panzer IV F2 vehicles were received mounting the long 7.5 cm KwK 40, a much more powerful weapon than the Sherman's 75.

The time was then considered ripe for planning a successor to the Sherman. Such a successor required improvements in firepower, mobility, and armor protection if it was to maintain an advantage over expected German developments. It was intended that the new series of tanks would take advantage of the lessons learned in battle as well as the technical advances that had occurred since the design of the Sherman. General Barnes and Colonel Colby proposed what was then considered a rather radical design to

An artist's conception of an early stage in the design study for a new tank. The sponsons are retained and the turret bears a strong resemblance to the Sherman. Note the sloped side armor.

meet these objectives. The new tank retained the same general layout as the Sherman with a five man crew. The driver and assistant driver were seated in the front hull while the tank commander, gunner, and loader rode in the turret. To achieve the maximum armor protection with a minimum of weight, a new technique of "Space Engineering" was used to design the hull. Using this concept, the hull was simplified to a box like structure eliminating the sponsons found on the M4. The simple shape reduced the surface area, allowing a given volume to be enclosed with thicker armor than the earlier more complex design, while maintaining the same total weight. The volume to be enclosed was also minimized by removing all nonessential equipment from inside the hull. Many of these items, carried inside the Sherman, were placed in external stowage boxes over the tracks. The hull was designed to fit the new low silhouette Ford V-8, Model GAN, tank engine, greatly reducing the overall vehicle height. The GAN engine was essentially the Model GAA used to power the medium tank M4A3, modified to fit in a low silhouette vehicle. At a later date, other minor modifications changed the designation to GAF. Addition of an automatic transmission also promised improvements in ease of handling and mobility.

The hull and fighting compartments were ventilated by a rotoclone blower mounted on the hull roof between the driver and assistant driver. The blower was intended to maintain a slight positive pressure within the vehicle when buttoned up and expel the powder gases when the guns were firing. The housing and intake for the blower formed a large bulge at the top center of the front armor plate giving a characteristic appearance to all tanks of the T20 series.

The proposed main armament was the new 76 mm gun T1, a light weight weapon developed to increase the firepower of medium tanks then equipped with the 75 mm gun M3. Although designated as a 76, the bore was actually 76.2 mm, or 3 inches in diameter and 57 calibers in length. After test, it was shortened to 52 calibers and standardized as the 76 mm gun M1. Further testing produced the M1A1 with changes in the external tube contour. The recoil surface was also lengthened by 12 inches permitting the trunnions to be mounted closer to the center of gravity giving better balance to the long gun. The new weapon was originally intended to fit into the Sherman turret using the same mount, but tests showed this arrangement to be overcrowded. To avoid a lengthy ammunition development program, the new 76 was designed to use the standard projectiles for the 3 inch gun already in service. These projectiles were fitted to a smaller cartridge case, but the powder charge was adjusted to obtain the same muzzle velocity as the 3 inch piece. With armor piercing ammunition, the muzzle velocity was 2600 ft/sec compared

to 2030 ft/sec for the Sherman's 75 mm M3 tank gun. The higher velocity increased the armor penetration by approximately one inch and was slightly superior to the German 7.5 cm KwK 40 in the Panzer IV F2. With a light weight tube and breech ring, the 76 was approximately two thirds the weight of the 3 inch tank gun M7.

In May 1942, a mock-up of the proposed tank was constructed by the Product Study Division of General Motors Corporation. The design offered a medium tank with ½ inch more armor, a 76 mm high velocity gun, and an automatic transmission for approximately the equivalent weight of the M4. Generals Devers and Somervell, as well as other high ranking officers, viewed the mock-up and gave their enthusiastic approval. Ordnance Committee action designated the new vehicle as the medium tank T20 and approved the construction of two pilot tanks.

Since the 76 mm gun was just being developed, alternate armament combinations were considered. These included the 3 inch gun M7 as mounted in the M6 heavy tank and the M10 tank destroyer, and a 75 mm M3 gun with an automatic loader. It was also desired to investigate the best possible combination of transmission and suspension systems. For this purpose four additional pilot vehicles were authorized in September, 1942. Two of these, designated medium tank T22, were to be equipped with a rearranged version of the Sherman's gear box transmission and were to be built at Chrysler Corporation. The other two vehicles were designated medium tanks T23 and used the electric drive developed at the General Electric Company for the T1E1 heavy tank. The following numbers were assigned for the various combinations of armament.

Medium tanks T20, T22, and T23 were to be equipped with the 76 mm gun M1. Medium tanks T20E1, T22E1, and T23E1 were to be fitted with the 75 mm gun M3 in a special turret with an automatic loader. The T20E2, T22E2, and T23E2 designations were reserved for vehicles armed with the 3 inch gun M7.

By the time the first pilots were constructed, the 76 mm gun had proved successful. Since the 3 inch gun was heavier in weight and identical in performance to the 76, no models of the T20E2, T22E2 or T23E2 were ever constructed. Only one turret was completed with the 75 mm gun M3 using the automatic loader. This was fitted to medium tank T22, pilot number 1 and the vehicle was redesignated medium tank T22E1. No T20E1 or T23E1 vehicles were built. The second pilot T20 was fitted with a torsion bar suspension and designated T20E3. As the development program progressed, additional numbers were assigned to cover various modifications of running gear, power trains, armament, and armor protection. These are discussed in detail in following sections.

Development Chronology of the T20 Series, 1942-1945

Year	Month	T20	T20E3	T22	T22E1	T23	T23E3	T25	T25E1	T26	T26E1	M45(T26E2)	M26(T26E3)	T26E4	T26E5
1942	April														
	May	Mockup completed													
	June														
	July														
	August														
	September														
	October														
	November				Gun mount proof fired										
	December														
1943	January			Pilots #1&2 completed		Pilot #1 completed									
	February														
	March					Pilot #2 completed									
	April														
	May	Pilot tank completed													
	June		Pilot tank completed *												
	July				Pilot tank converted										
	August														
	September														
	October					Production started									
	November														
	December														
1944	January							Pilot #1 completed	Production started		Production started				
	February														
	March														
	April							Pilot #2 completed	Production of 40 tanks completed		Production of 10 tanks completed				
	May														
	June														
	July														
	August						Pilot tank completed								
	September														
	October									Pilot tank completed			Production started		
	November														
	December					Production of 250 tanks completed									
1945	January													1st temp. pilot completed	
	February												Committed to battle		
	March												Standardized as M26		
	April														
	May														
	June											Production started			Production started
	July														
	August														
	September														

* Standardization proposed as M27 (T23E3) and M27B1 (T20E3)

The T20 pilot after completion at the Fisher Body Division of General Motors in May 1943. The lugs on the right side of the turret were for mounting an early version of the hoisting device used for removing the power train.

THE T20 AND T20E3

In May 1942 the contract for the T20 pilot tanks was given to the Fisher Body Division of General Motors Corporation and the first T20 pilot was completed a year later in late May 1943. The vehicle reflected the design changes resulting from the January conference between Ordnance and the Armored Force. These changes improved the sitting height for the drivers and enlarged their entrance hatches. The new front end was a welded assembly of castings and rolled plate. The same conference also concluded that the tanks would be approximately 2½ tons heavier than the original design weight and that numerous improvements were necessary, particularly in the ammunition stowage. The quantity of ammunition carried was considered below the minimum required and insufficient protection was provided for the rounds stowed in the turret. This criticism also applied to the other T20 series pilots under construction at that time. However, it was agreed that the pilots would be completed for comparison test purposes.

T20 number 1 was equipped with an early type of horizontal volute spring suspension using the same 16 9/16 inch wide double pin, rubber block track as the medium tank M4. Two wheel bogies were mounted, three on each side of the vehicle, with

hydraulic direct acting shock absorbers controlling the movement of the front and rear suspension units. M4 type track support rollers were attached directly to the hull at the rear of each bogie and track tension could be adjusted by moving the idlers fitted at the front corners of the hull. The tracks were driven by sprockets attached to the final drives mounted on each side at the rear. Power from the Ford tank engine passed through the new type torqmatic transmission to the differential and then to the final drives. The torqmatic transmission was the Model 30-30B, consisting of an hydraulic torque converter in series with planetary gearing allowing three speeds forward and one reverse. The complete power train consisting of engine, transmission and differential was removable as a unit.

The registration number for the first pilot was 30103302. After completion, the tank was sent to the General Motors Proving Ground for tests and then returned to the manufacturer for correction of defects. The torqmatic transmission proved to be a constant source of trouble. The major defects were oil leaks and overheating. The T20 was shipped to Aberdeen Proving Ground in February 1944. However, more promising developments were now available and no further tests were conducted.

The early type horizontal volute spring suspension can be seen on these views of the T20 pilot. Close examination of the side view will show the shock absorbers mounted on the front and rear bogies. The front view shows the weld joints between the upper hull casting and the rolled front plate. The joints are also visible around the ball mount casting and the adjustable idler mounts at each front corner.

The T20 pilot is shown above with the gun forward (top left) and locked in the travel position (top right). The circular rotating hatch for the tank commander is the same as on the M4 medium tank with a mount for an antiaircraft machine gun and a periscope in one side of the split cover. This early version of the loader's hatch with the rectangular split cover is oriented parallel to the axis of the gun tube. The loader's periscope cover is just forward of the hatch. The tank is shown with all covers open at the left. The vertical mounted cooling fans can be seen at the rear of the engine compartment.

The T20 pilot under test at General Motors Proving Ground in June 1943. The gun is locked in the travel position in both views. Note the absence of a pistol port in this turret.

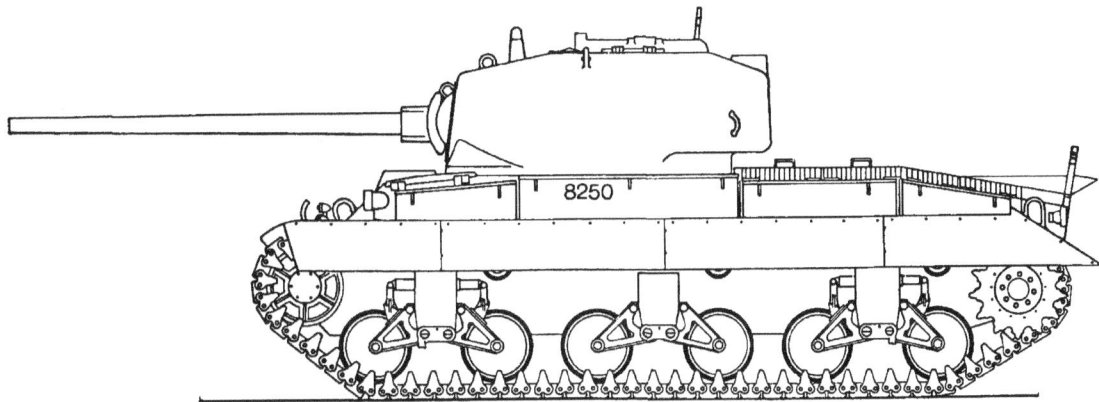

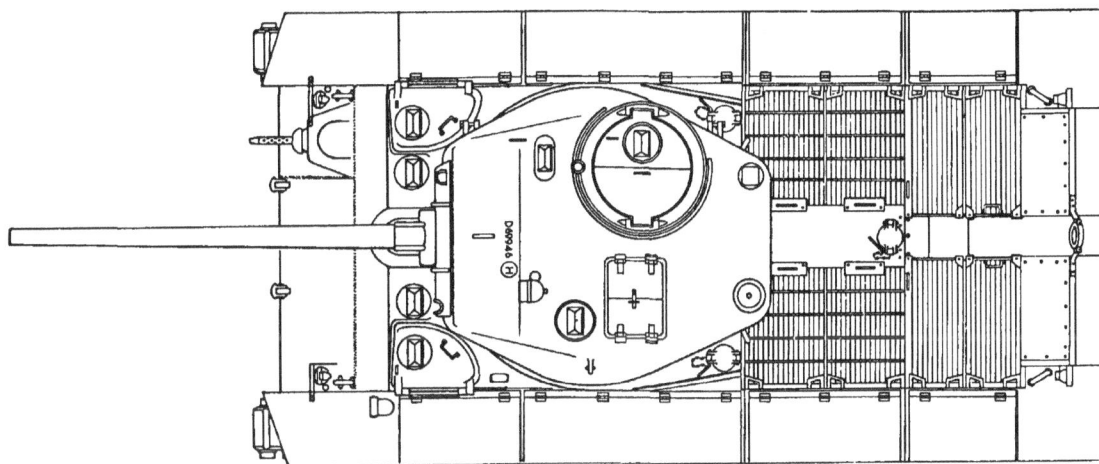

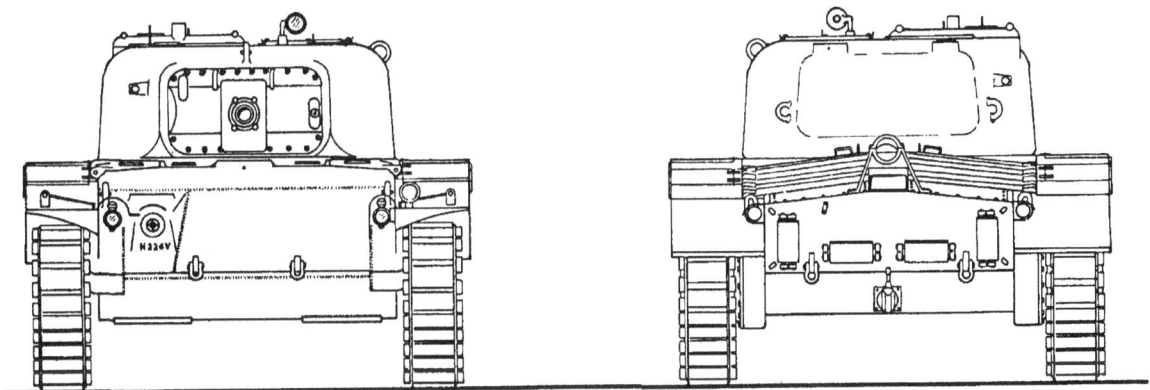

8250

Scale 1:48

Medium Tank T20, Pilot

The breech of the 76 mm gun can be seen at the bottom of the picture at the left above. The weapon is mounted so the breechblock slides in the horizontal direction. The empty bracket for the coaxial .30 caliber machine gun can be seen just to the left of the 76. The hole for the gunner's telescopic sight appears to the right of the gun mount. At the right above is a view looking into the turret through the tank commander's hatch. The upper and lower seats for the tank commander are in the folded position. The recoil guard for the 76 mm gun appears at the left side of the opening.

The driver's and assistant driver's positions are shown below at the left and right respectively. The instrument panel can be partially seen mounted between the two drivers and the empty bow machine gun mount appears at the right. The bolt heads in the side walls are for mounting the forward suspension bogies.

DRIVER'S COMPARTMENT

ASSISTANT DRIVER'S COMPARTMENT

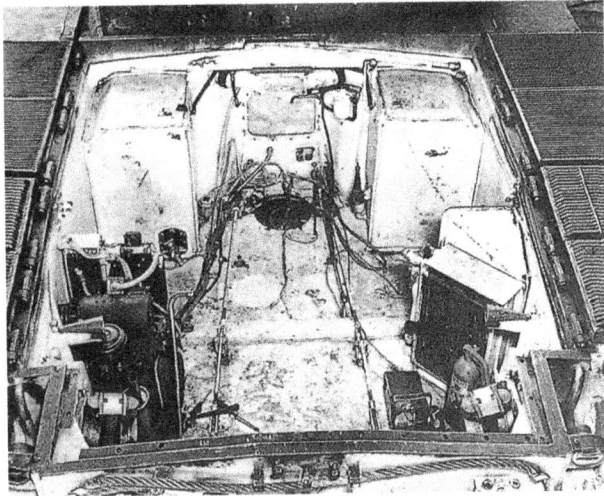

COOLING FANS

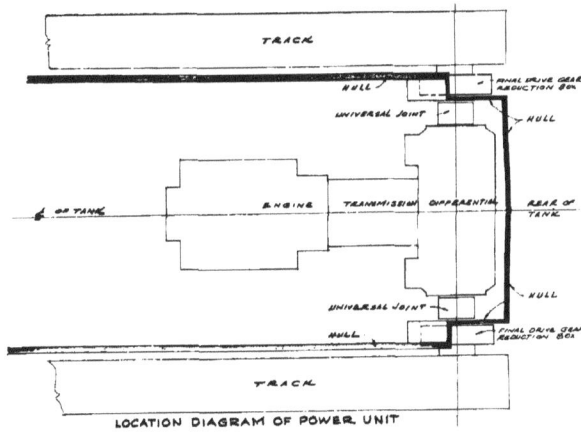

TRACK

FINAL DRIVE GEAR REDUCTION BOX

HULL

UNIVERSAL JOINT

HULL

OIL TANK

ENGINE   TRANSMISSION   DIFFERENTIAL   REAR OF TANK

HULL

UNIVERSAL JOINT

HULL

FINAL DRIVE GEAR REDUCTION BOX

TRACK

LOCATION DIAGRAM OF POWER UNIT

The engine compartment is shown (top left) with the power unit removed. Fuel tanks are located in the forward end with the auxiliary engine and generator installed along the left wall. Control rods extend along the bottom of the compartment. The power unit is installed (top right) and the radiators can be seen behind the vertical cooling fans. The location of the various power unit components is diagramed at the left.

The power unit removed from the tank appears in these photographs. A closeup (at left) shows the rear of the unit and the controlled differential. With this arrangement the cooling fans were removed with the power unit. The linkage required to drive these fans can be seen in the photograph at the right.

POWER UNIT

The Armored Force in September 1942 recommended that one of the T20 pilots be completed with a torsion bar suspension and wider tracks in an effort to improve the ride and reduce the ground pressure. To avoid confusion, the second pilot was redesignated T20E3 by Ordnance Committee action in February 1943. The pilot T20E3, with registration number 30104303, was completed on July 1, 1943. This vehicle was equipped with a torsion bar suspension which was a modified version of that just going into production on the 76 mm gun motor carriage M18. As fitted to the T20E3, the suspension consisted of six dual disc road wheels on each side of the vehicle. These wheels were mounted on support arms attached to the ends of torsion bars fitted transversely across the bottom of the tank. This arrangement allowed a wheel movement up to about eight inches and shock absorbers were applied to the first two and the last two wheels on each side. The idler wheel was connected to the front road wheel to compensate for slack in the track. As originally constructed, three pairs of dual track support rollers were fitted. Later, an additional two pairs were added, giving five dual track support rollers on each side of the vehicle. An 18 inch wide single pin, cast steel, center guided track was driven through extensions of the track pins.

Although completed on July 1, 1943, the T20E3 was retained temporarily by the manufacturer for the correction of defects. As a result of a fire, a large part of the vehicle had to be rebuilt, using some parts from the T20 pilot. The tank was then shipped to Aberdeen Proving Ground for comparative ride

tests with M4 medium tanks equipped with the vertical volute spring suspension. These tests indicated a considerable ride improvement over the vertical volute spring suspension, but showed the need for strengthening the shock absorbers and their mountings. Transmission difficulties similar to those encountered with the T20 pilot precluded extensive tests and the Ordnance Committee recommended that the project be cancelled in December 1944.

A design study in April 1943 showed the T20 equipped with the original version of the General Motors cross-drive transmission. An extremely compact unit, the new transmission was actually tested in a modified M4A3. After further development, the cross-drive was applied to the M26E2 and then to the M46 during the postwar period.

The turrets of both the T20 and T20E3 were armed with the 76 mm gun and a .30 cal machine gun in the coaxial combination gun mount T79. This was a modified version of the Sherman's 75 mm gun mount M34. Both vehicles were equipped with essentially identical cast turrets reflecting the stage of turret development at the time of their construction. The tank commander was provided with a revolving ring type hatch with a fitting for a .50 cal antiaircraft machine gun. The loader's hatch was smaller, rectangular in shape, with double doors. The turrets were obviously produced during the period when pistol ports were out of favor as neither was fitted with one. The discussion of turret and gun mount development in the section on the T23 applies equally to other models of the T20 series.

Pilot T20E3 after completion at Fisher in July 1943. Note there are only three track return rollers per side.

The single pin, center guided tracks and torsion bar suspension show a strong resemblance to those on the 76 mm gun motor carriage M18 from which they were developed. A 27 tooth sprocket replaced the familiar 13 tooth sprocket found on the T20. The shock absorbers on the first two and last two road wheels are visible in the upper and lower photographs. The gun traveling lock of the T20 has been replaced with a single rest for the gun tube and a new more rounded casting design has been applied to the bow machine gun mount.

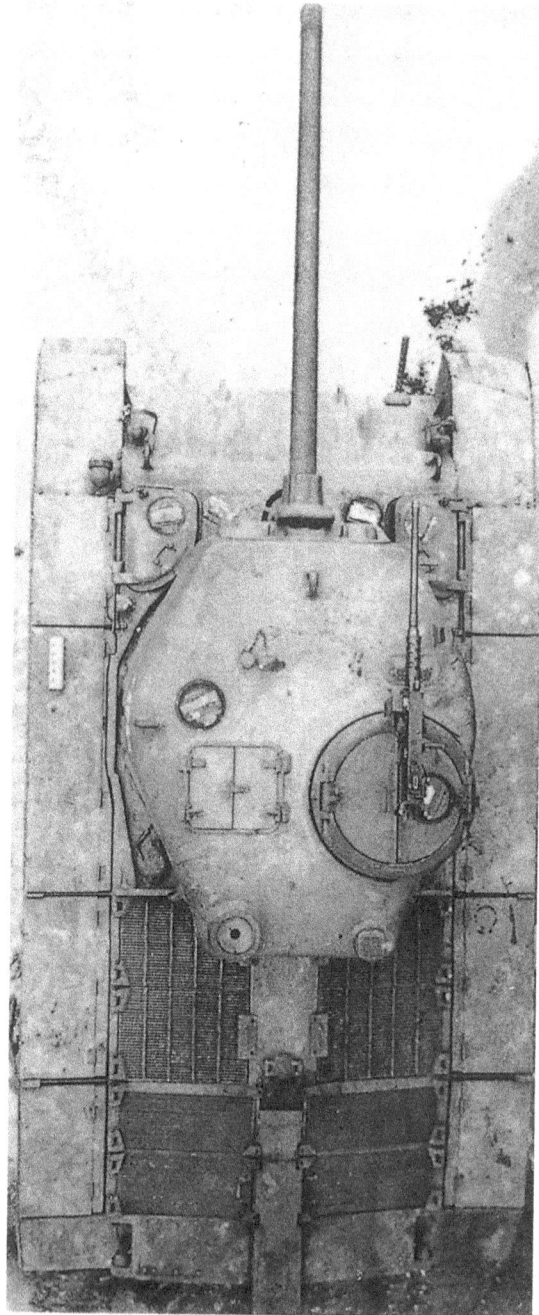

Later photographs of the T20E3 at Aberdeen Proving Ground show the modified suspension with five track return rollers. The .30 and .50 caliber machine guns have also been installed.

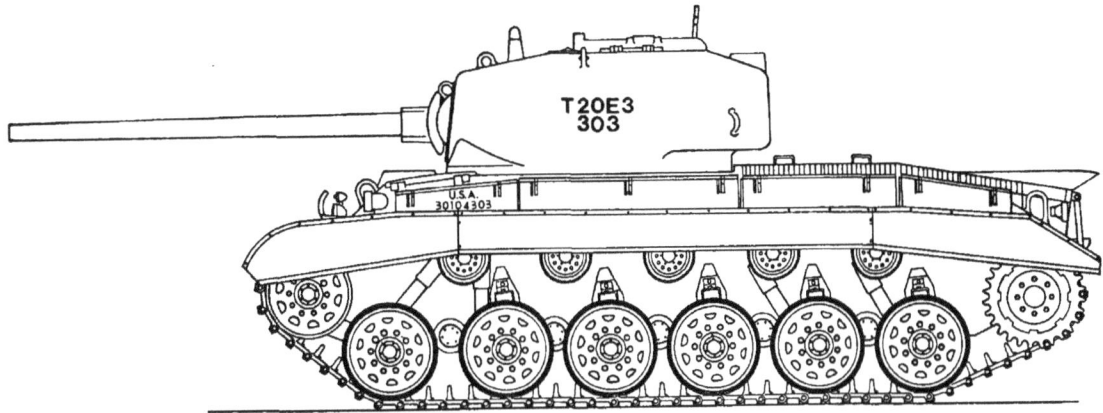

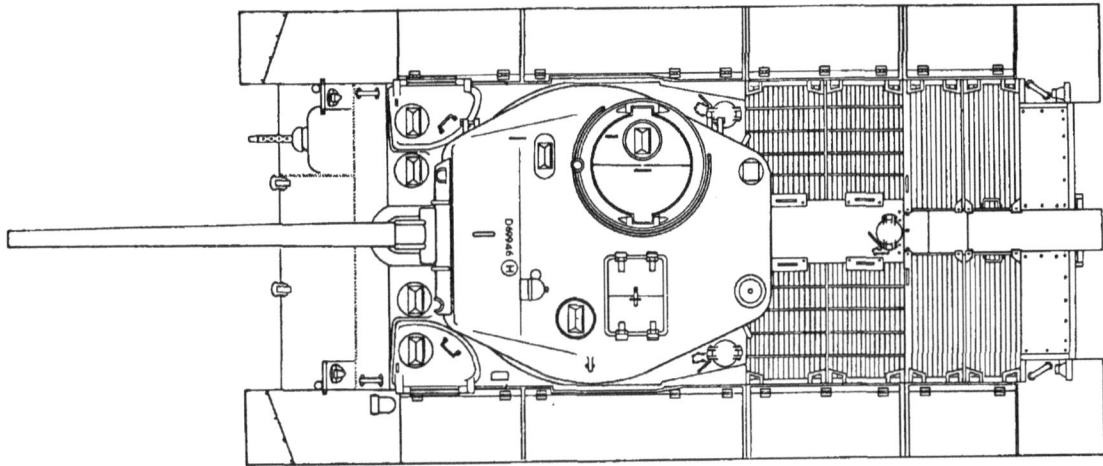

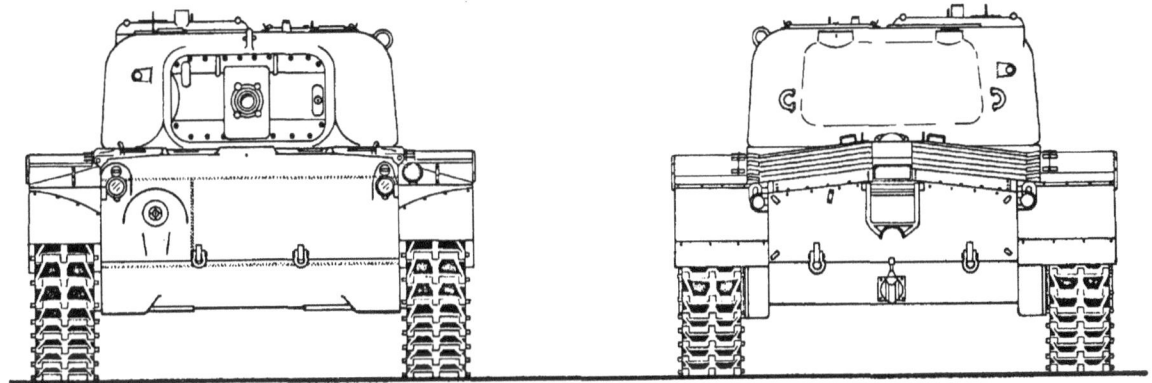

Scale 1:48

Medium Tank T20E3, Pilot

T22 pilot number 1 after completion at Chrysler in June 1943. The shock absorbers on the front and rear bogies of the early horizontal volute spring suspension are easily seen in this photograph.

THE T22 AND T22E1

The two pilot medium tanks T22 were constructed by Chrysler Corporation and completed in June 1943. Their registration numbers were pilot number 1, 30104304 and pilot number 2, 30104305. These tanks followed the same basic design as the medium tank T20. However, they were fitted with the power train components of the medium tank M4A3, but these components were modified and rearranged. For example, the transmission and differential, although similar to those used in the M4 series vehicles, were mounted in different type cases and the assembly operated as a dry sump, with no gears dipping in oil. The gear box transmission and controlled differential were fitted behind the engine in the hull. The transmission and differential had the same gear ratios as the M4A3 with five speeds forward and one reverse. As with the T20, the final drives were mounted on the outside of the hull at the rear of the tank. The early horizontal volute spring suspension with direct acting shock absorbers on the front and rear bogies was applied to the T22s as well

as the T20. This type of suspension gave some improvement in ride over the Sherman's vertical volute spring system. However, in this early form with the same narrow 16-9/16 inch track, the improvement was insufficient to warrant slowing production of the Sherman to make a change.

Other modifications were made based on the experience with the T20 pilot. To permit the circulation of cool air around the auxiliary generator, it was shifted forward about eighteen inches and the oil coolers were moved to the right side of the engine compartment. The batteries were transferred from the engine compartment to a position between the driver and assistant driver.

During tests at the Tank Arsenal Proving Ground, engine trouble was encountered in the first pilot after driving 900 miles. At this point, a connecting rod went completely through the crankcase wall requiring replacement of the engine. The second pilot vehicle was sent to Aberdeen Proving Ground where considerable engine and transmission difficulties occurred requiring the installation of a new power train. Because of these problems, further development work was discontinued and in February 1944 the tests were suspended because of higher priority work.

The left rear view with the gun in the traveling position shows the gun rest similar to that on the T20E3. T48, double pin, rubber chevron tracks are fitted to this tank.

Other views of T22 pilot number 1 are shown above and below. The turret and gun mount were quite similar to the T20 and T20E3, but the bow machine gun mount casting more closely resembles that of the T20. Note the pistol port has been welded over.

The T22 pilot number 2 under test at Aberdeen Proving Ground. Note that the welding of the upper front armor plate on both T22 pilots differs from the T20 and T20E3. The periscope in the commander's hatch cover is centrally mounted, unlike other hatches of this type. The periscope mount is fitted to one side of the split hatch cover extending into the simicircular opening on the other side when the hatch is closed. Compare with the top views of T22 number 1 and the T20.

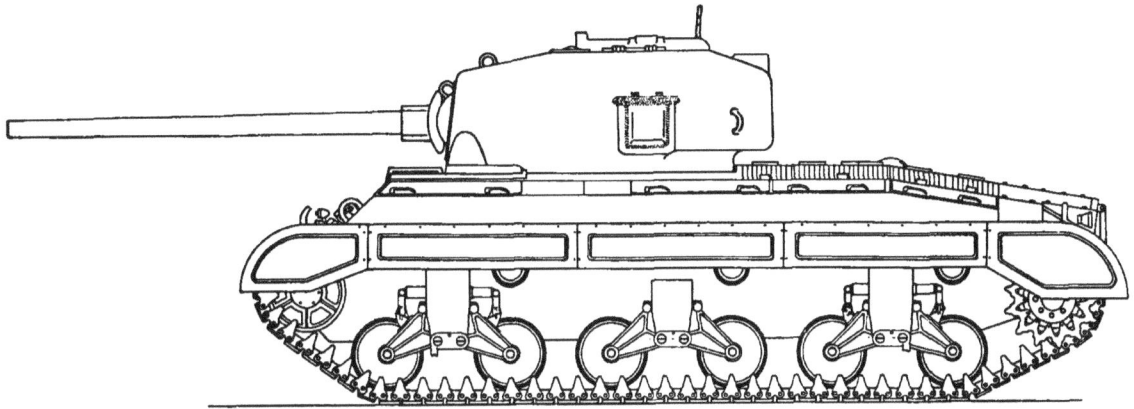

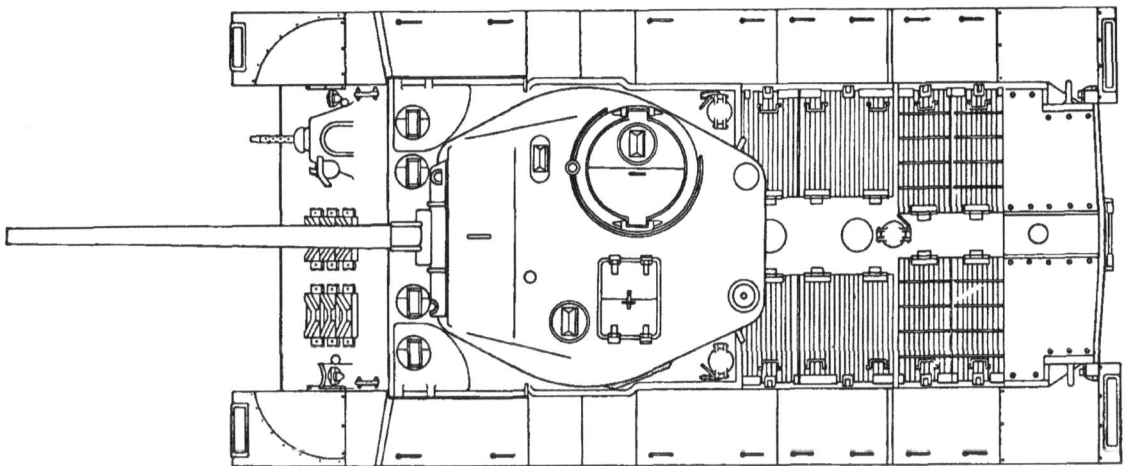

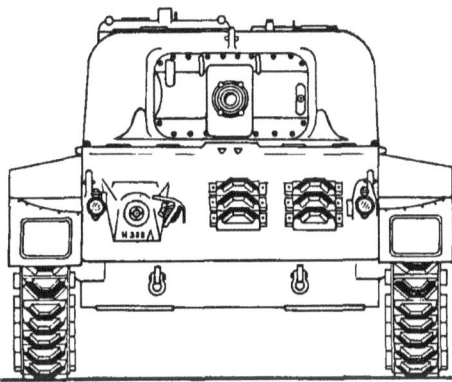

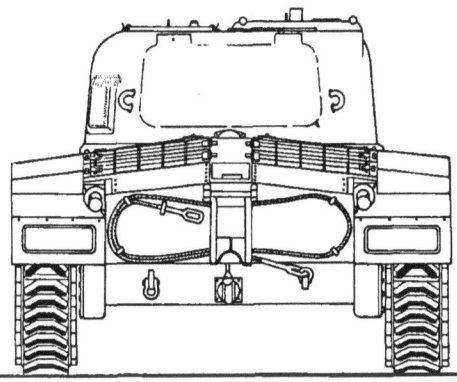

Scale 1:48

Medium Tank T22, Pilot Number 1

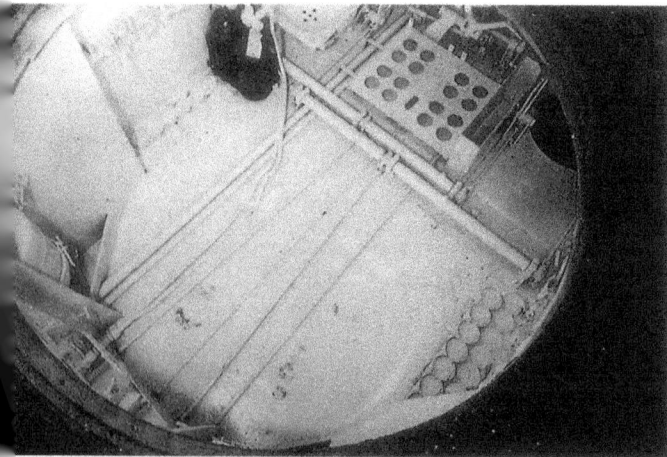

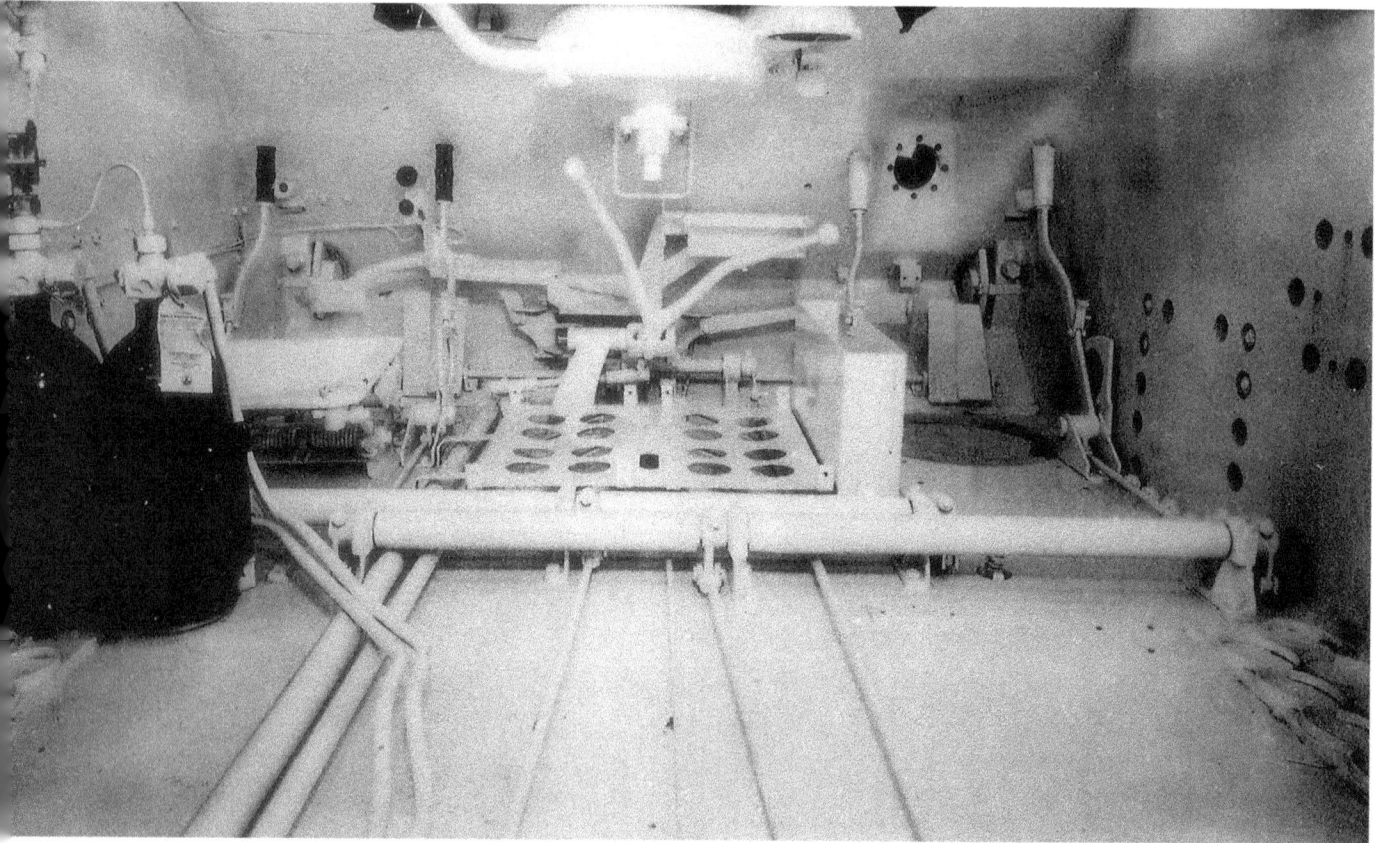

The drivers' compartment during assembly (above) shows the control linkage. The driver's seat is mounted, but the assistant driver's seat and escape hatch cover is still missing. The roto-clone blower appears at the upper center of the photograph.

The two photographs at the left are taken through the turret ring opening before installation of the turret. Looking forward (upper), the driver's seat and the fixed fire extinguisher bottles are visible. Looking toward the rear (lower) reveals the opening into the engine compartment with air cleaners on either side. Racks for 76 mm ammunition stowage are located on both sides of the hull.

The view above is looking toward the rear over the engine compartment prior to installation of the covers or the turret. The controls just below the turret ring are for the auxiliary engine and generator.

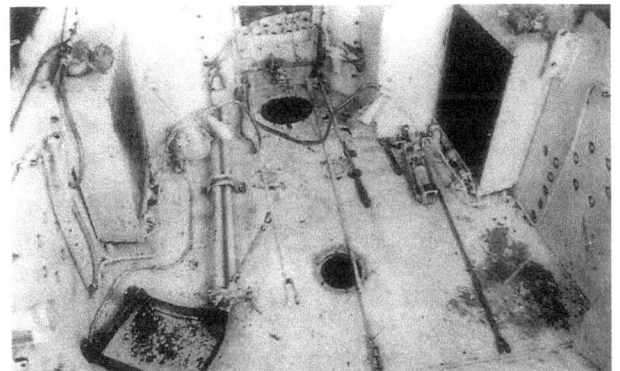

Two views above show the engine compartment with the power train removed. The fuel tanks appear on each side of the opening into the fighting compartment. The small conical jets are fire extinguisher outlets. The two views below are of the power unit with the fans removed.

Above and below are two views of the T22E1 pilot after conversion from T22 pilot number 1.

Both pilots initially carried the same type of cast turret and T79 gun mount as the T20. By this time, this turret design was considered obsolete and the 76 mm gun tubes installed in the pilots were unsafe for use. To prevent any firing, the tubes were plugged with wooden blocks. T22 pilot number 1 was later fitted with the special turret mounting the 75 mm gun M3 equipped with an automatic loader. This turret, designed and constructed by United Shoe Machinery Corporation, was originally intended for the medium tank T20E1. When fitted with this turret, the pilot was redesignated medium tank T22E1. The gun and mount had been tested by proof firing in November 1942. These tests revealed some high friction in the system and later tests in April 1943 revealed difficulty with the automatic spring rammer. After redesign it was fired automatically at a rate of 20 rounds per minute. The turret was fitted to the T22 vehicle in August 1943 and proof fired at Salisbury Beach, New Hampshire. As a result of these tests, a safety device to stop the loader, when an empty case was not completely ejected, was built into the mechanism and further firing tests were conducted.

The T22E1 turret was similar in size and shape to the standard 75 mm gun turret used on the medium tank M4. It also used the M34 gun mount employed on the M4 series. Since it was fitted with an automatic loader, the turret crew was reduced to two, the tank commander and the gunner. The tank commander was moved to the left of the 75 mm gun below a revolving hatch fitted with a .50 caliber machine gun for antiaircraft use. The gunner's position remained to the right of the 75 mm gun with a small escape hatch in the turret roof. The turret could be readily identified by the protrusion in the center rear of the roof necessary to permit clearance for the automatic loading mechanism. The loader itself was fed from two 32 round magazines, one on each side of the gun. Normally this meant 32 rounds of high explosive in one magazine and 32 rounds of armor piercing ammunition in the other. The selection of ammunition was made by the tank commander, using the control mounted on the left side of the turret.

The T22E1 was shipped to Aberdeen Proving Ground in November 1943 for complete firing tests. These tests indicated that an automatic gun gave superior performance to a hand loaded gun, but the equipment did not have sufficient mechanical reliability. In addition, the 75 mm gun no longer provided adequate firepower for a medium tank. For these reasons, the work on the project was suspended in February 1944.

T22E1 pilot under test at Aberdeen Proving Ground. A counterweight has been added to the gun muzzle to balance part of the automatic loading mechanism and improve the stabilizer performance. The bulge at the center rear of the turret roof is an immediate point of recognition. This was necessary to permit adequate clearance for the automatic loader.

RAMMER TRIP LEVER
SPRING RAMMER
BREECH
LOADER
ROUND
LOCATION OF HYDRAULIC UNITS
L. H. MAGAZINE
R. H. MAGAZINE
TRANSFER MECHANISM
LOADER TRAY

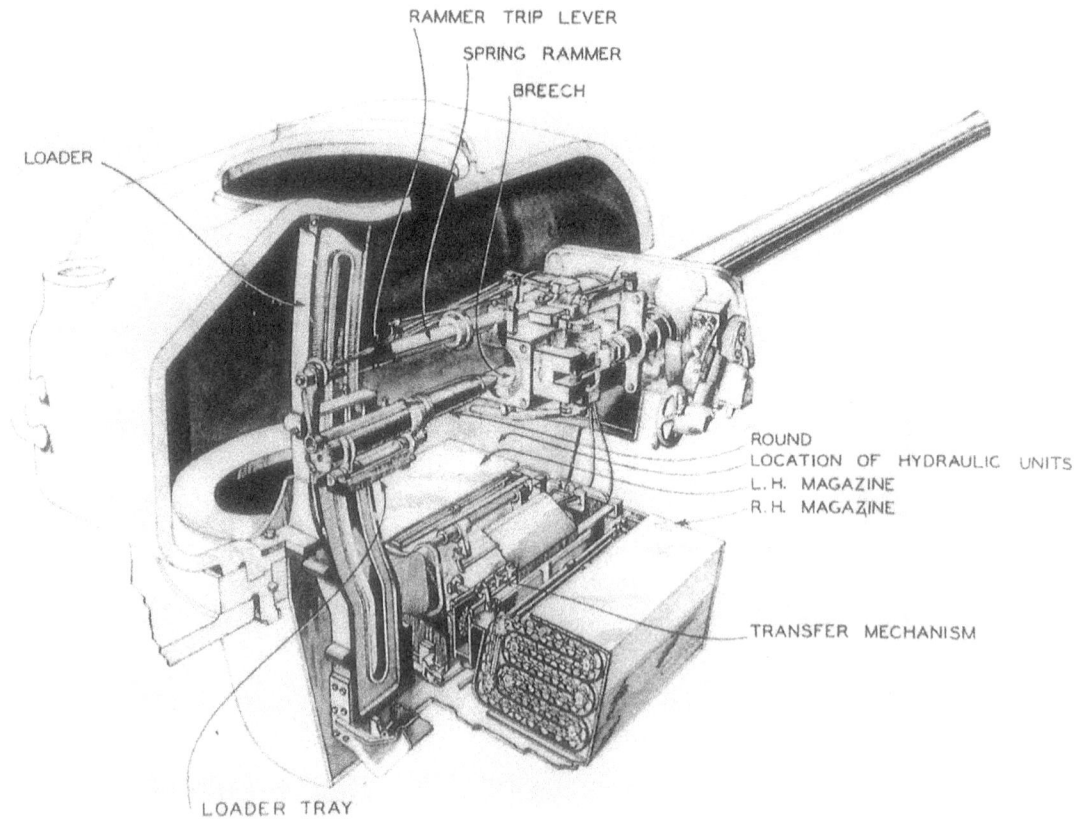

Sectional view of turret showing automatic loader

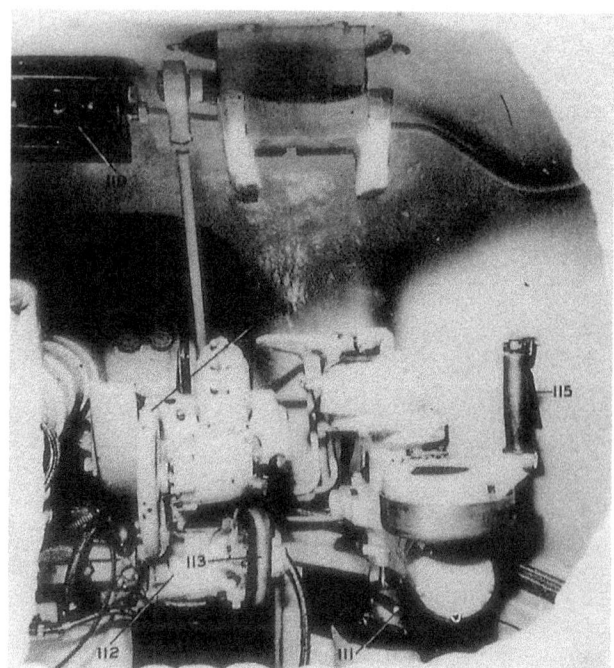

Interior views of the turret show the tank commander's position (above) and the gunner's station (right).

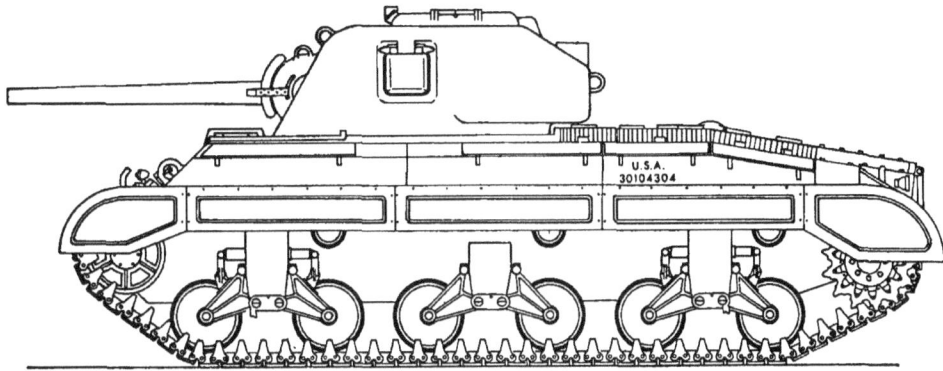

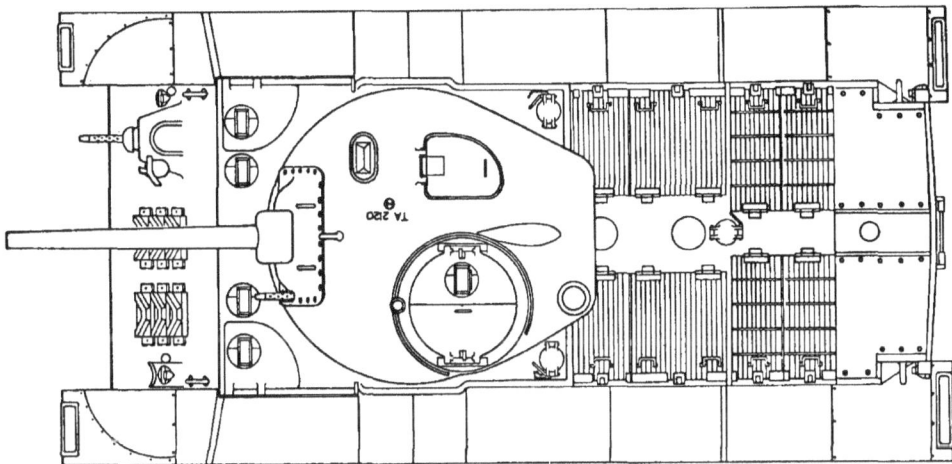

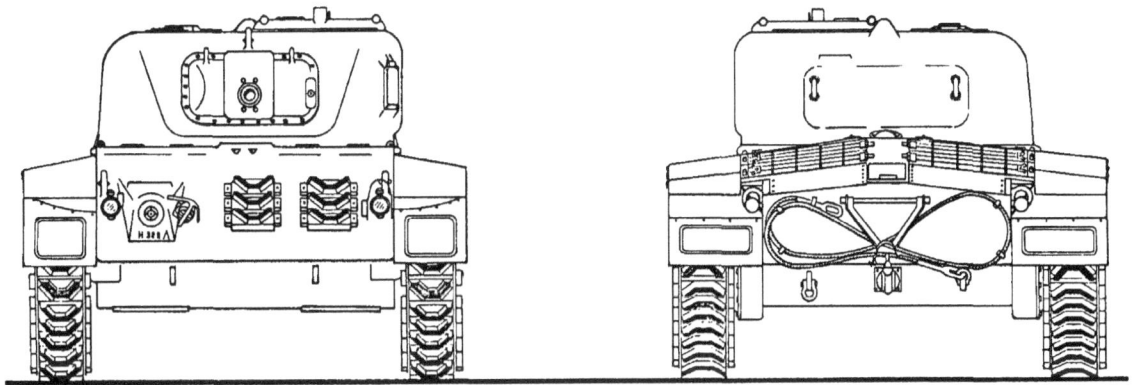

U.S.A.
30104304

TA 22O

Scale 1:48

Medium Tank T22E1, Pilot (converted from T22, Pilot Number 1)

71

T23 pilot number 2 in March 1943.

THE T23, T23E3, AND T23E4

Although highest in number, the first of the original pilots to be completed was the medium tank T23. Pilot number 1 (registration number 3098787), with the exception of the fighting compartment, was completed at the General Electric Company's Erie Works in January 1943. Construction had been authorized after the exceptional performance obtainable with an electric drive was noted during tests of the heavy tank T1E1. The same drive was modified for use with the Ford GAN tank engine in medium tank T23. With this system, the engine drove an electric generator, the controlled output of which powered two individual traction motors operating the final drives. Since there was no mechanical connection between the engine and tracks, the engine speed could be varied and controlled without being in direct proportion to the vehicle speed. It was therefore possible to operate the engine at its most efficient speed at all times. Such a design held the possibility for an appreciable increase in engine life. Preliminary tests showed the tank was extremely maneuverable, could turn in its own length, and had a maximum speed of 35 miles per hour. Unlike the other T20 series pilots, the T23 was fitted with the same vertical volute spring suspension found on the M4 tanks. The sprockets were mounted on the final drives at the rear of the hull and track tension was maintained by adjustable fixed idlers at the front of each track. The suspension carried the standard 16-9/16 wide T51 track.

Pilot number 1 was completed before the decisions reached at the January meeting between Ordnance and the Armored Force were put into effect.

However, the hull for the second pilot was refabricated to increase the sitting height and enlarge the drivers' hatches. The upper front hull of both vehicles was welded from rolled plate fitted around the rotoclone housing. The large casting including the rotoclone housing and extending across the full width of the hull was not introduced in the T23 until the production pilot was completed in October 1943.

The T23 pilot number 1 also differed from the other pilot vehicles in being fitted with a welded turret. This resulted from the program to develop an improved turret and mount for the 76 mm gun. The effort began during the months of July and August 1942 when the original 76 mm gun T1 was installed in the combination mount M34 and fitted to a medium tank M4A1. The new installation required a counterweight of approximately 800 pounds to balance the gun mount. The elevating mechanism was too weak and the turret was cramped, preventing efficient operation of the gun and mount. To correct this, a design study was initiated to develop an improved turret for mounting a 76 mm gun on medium tanks. The 76 had meanwhile been shortened five calibers or 15 inches and modified to improve its balance. It was around this weapon, the 76 mm gun M1A1, that the design proceeded. The 76 was equipped with an elevation stabilizer in all T20 series vehicles.

After preliminary studies, it was decided to proceed with the detailed design of two different turrets and mounts. One gun mount design was based on the combination mount M34, already in use with the 75 mm gun in the M4 tanks. The gun cradle was a rotor

T23 pilot number 1 with welded turret and the early T80 gun mount. Note the vertical volute spring suspension as on the Sherman.

mounted on bearings in a fixed front plate which was bolted to the machined surface on the front of the turret. A small rotor shield was fitted around the gun barrel and bolted to the rotor to protect the machined surface exposed through the opening in the fixed front plate when the gun was elevated or depressed. A direct sight telescope was added for the gunner as well as the standard M4 periscope sight. A coaxial .30 caliber machine gun was mounted to the left of the main weapon. This version was assigned the designation, combination gun mount T79 and the turret to which it was assembled was based on the standard M4 design. The turret was of cast construction and, in addition to the revolving hatch over the commander's position, provided a smaller hatch with double doors for the loader. The commander and loader were each equipped with an M6 periscope and the revolving ring on the commander's hatch mounted a .50 caliber antiaircraft machine gun. Photographs taken in March 1943 show this turret on the second pilot T23 (registration number 3098788) and the same type was installed on the T20, T20E3, and the T22.

In the second type of gun mount, the turret front plate was eliminated and replaced by a large moveable shield to which the gun cradle was bolted. This shield was attached to the turret body by means of trunnion bearings, the trunnion pins which carried these bearings being supported by the turret body. The design was assigned the designation, combination gun mount T80, and the original turret body to which it was assembled was of all welded construction. This turret was much better balanced than the cast turret used with the T79 mount, since the bustle extended further to the rear.

A single turret basket design was used for both turrets. Forty-two rounds of 76 mm ammunition were carried in this basket as well as sufficient caliber .30 and caliber .50 ammunition for the machine guns. It was the welded turret with early T80 gun mount which was fitted to medium tank T23 pilot number 1.

T23 pilot number 1 during tests at General Electric to determine the center of gravity.

After satisfactory results were obtained in tests, the combination gun mount T80 was released for production. The welded turret for this mount was redesigned as a cast turret and the bulge contour at the rear changed to improve the balance and permit its use on the M4 tank series as well. The change was necessary to clear the drivers' hatches on the M4 tanks. When mounted on the M4 series, it was also necessary to provide a ventilating fan at the rear of the turret, since those vehicles were not equipped with a hull mounted rotoclone blower. This was the turret adopted for the production T23 tanks with the T80 mount standardized as the M62.

Details of the welded turret are revealed in this example prepared for ballistic tests. It has the same general arrangement as the cast turret.
Note that the pistol port has been welded shut.
A view of the turret roof (below) during tests at General Electric shows details of the hatches and periscope covers. The seat was obviously mounted for use during test operations.

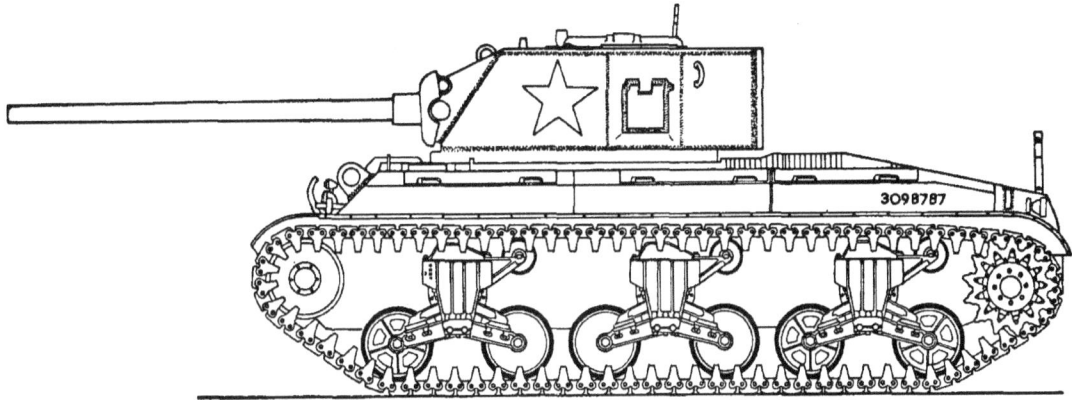

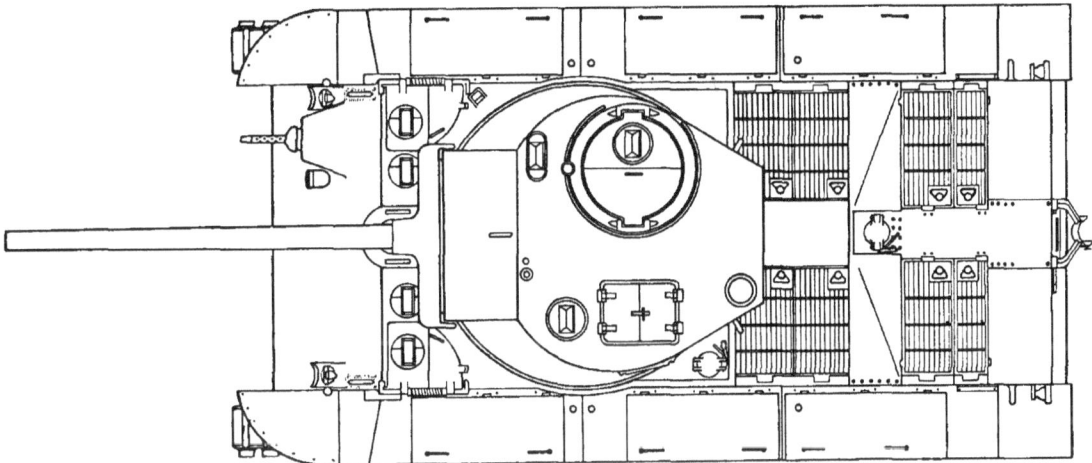

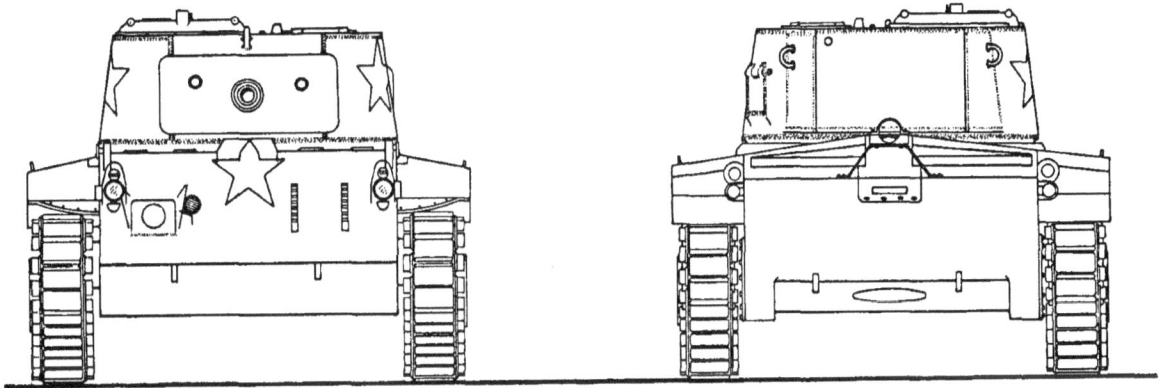

Scale 1:48

Medium Tank T23, Pilot Number 1

T23 pilot number 2 in March 1943 with the cast turret and T79 gun mount. The pistol port has been eliminated on this model and no traveling lock is fitted for the 76 mm gun.

Right Track
Driving Motor   Resistor

Left Track
Driving
Motor

Generator

Amplidyne
Generator
Assemblies

Ford Model
Gan Engine

Further details of pilot number 2. The power train arrangement is shown above. Note also, that a .30 caliber machine gun is mounted on the tank commander's hatch. Later, a .50 caliber machine gun was standard for this position. The engine exhaust is the opening between the spare track links in the rear view. Periscopes for the drivers were mounted both in the hatch covers and adjacent to the hatch (below left). This view also shows the welded assembly of the front hull armor.

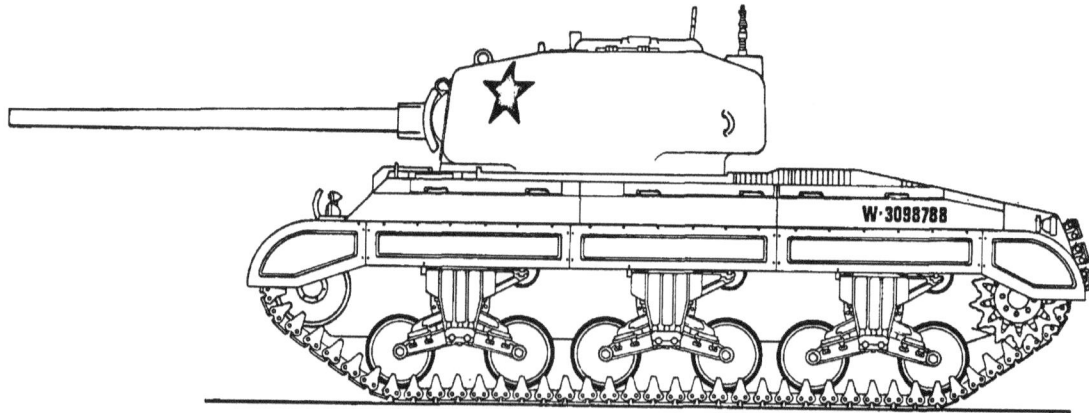

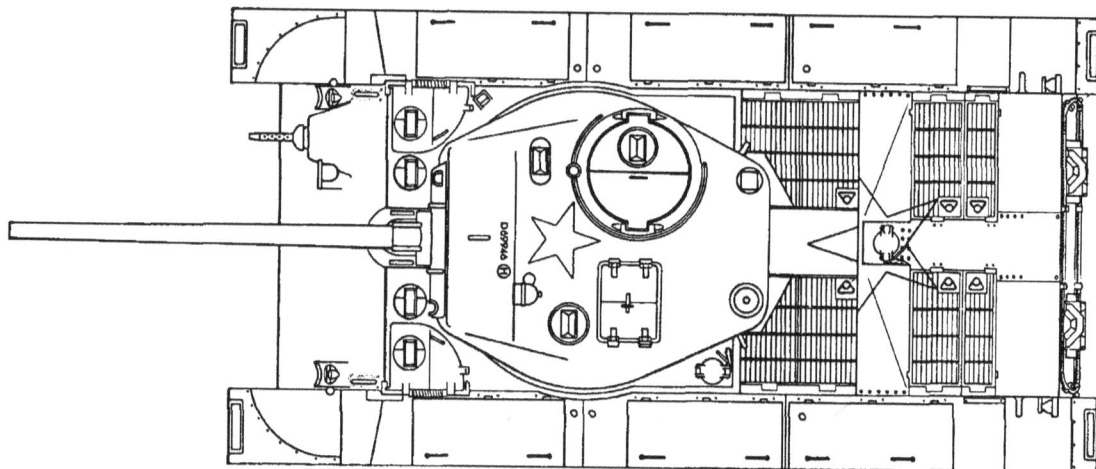

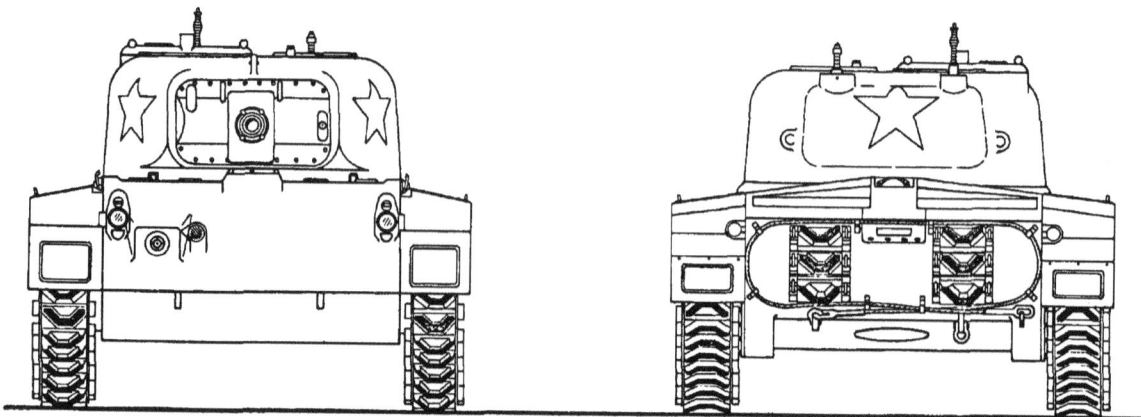

Scale 1:48

Medium Tank T23, Pilot Number 2

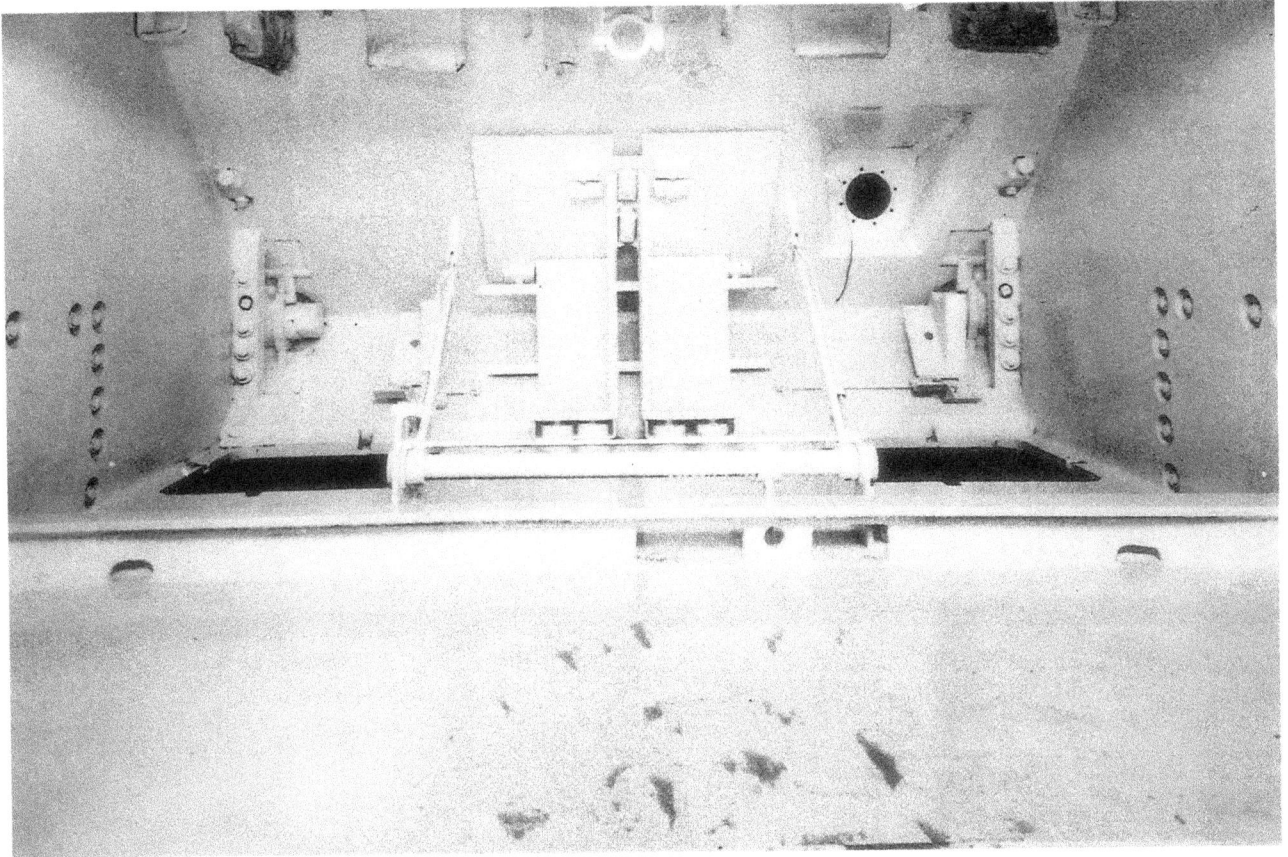

The drivers' compartment during assembly (above) compared with each position below after stowage was complete. At left below, the driver's controls are seen at each arm of his seat. Similar controls are provided for the assistant driver (below right) plus the .30 caliber machine gun in the bow mount. The escape hatches are in the floor directly in front of each seat.

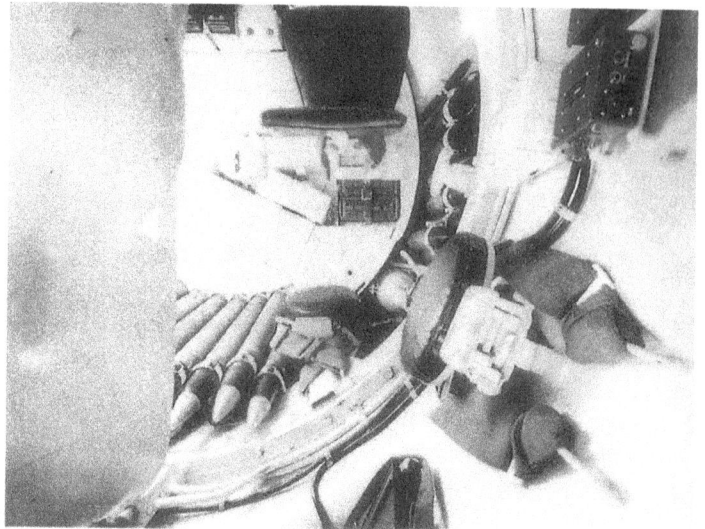

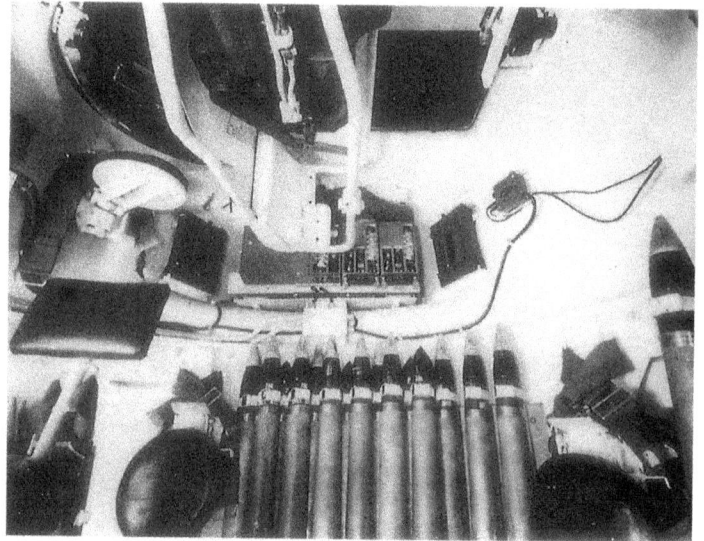

Turret stowage arrangements in pilot number 2 are shown here. The loader's position (above) provides ready access to the 76 mm ammunition stowed around the turret. Note the counterweight in the recoil guard necessary to balance the 76 mm gun. A view through the commander's hatch (top right) shows his upper and lower seats folded. The gunner's seat is at the top center of the photograph. Looking toward the rear of the turret (center right), the radio can be seen mounted in the turret bustle partially obscured by the recoil guard of the 76 mm gun. The remaining view (bottom right) is from underneath the gun and shows the access to the driving compartment. The gunner's seat is at the right.

The production pilot T23 (serial number 3). Note the welded assembly of castings and rolled plate in the front hull. The pistol port is still absent from this first production tank, but a traveling lock has been added for the gun.

Details of the new production turret are shown below. The tank commander has been equipped with a vision cupola and the rotating hatch moved to the loader's position. Lugs for mounting a power train hoisting device are welded to the right side of the turret.

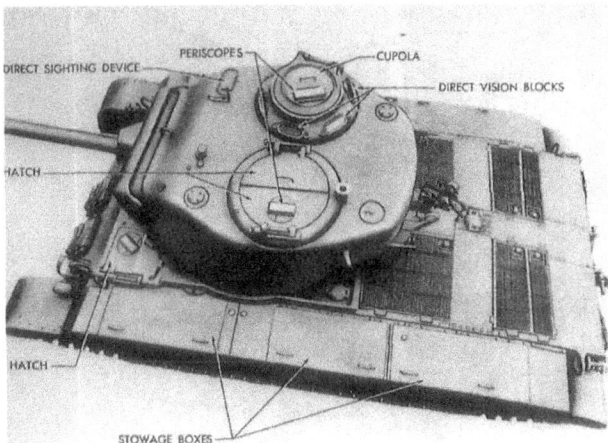

The preliminary tests on the T23 pilot models proved so promising that a production contract for 250 tanks was approved in May 1943. The registration numbers were 30103052 through 30103301. However, 30103252 through 30103301 were later assigned to the T25E1 and T26E1 and the number of T23s reduced to 200 on this contract. A later contract for an additional 50 T23s brought the total back to 250 by the time production ceased in December 1944. The production total of 250 also included the two T25 pilots originally classified as T23s with the 90 mm gun. Their registration numbers were 30103053 and 30103054. The first production pilot incorporating several modifications was completed in October 1943 and shipped to Aberdeen Proving Ground.

The production T23 carried the improved cast turret with a T80 (M62) gun mount and was fitted with a vision cupola for the tank commander. The revolving hatch with the antiaircraft machine gun was moved to the loader's position and a 2 inch smoke mortar M3 was mounted in the left front turret roof. The front armor plate thickness was increased to 3 inches instead of the 2½ inches on the pilots. Additional improvements based on tests at Fort Knox were incorporated during the production run. The later production vehicles had complete water proofing for all electrical parts and electric brakes for emergency stopping, if power was lost. Improved engine components such as valves, gaskets, and spark advance governors were also included as a result of the testing program.

An interesting feature of the electric drive tanks was their ability to be driven by remote control from the turret or by carrying the remote controller while walking alongside the vehicle. For operating over the full range of available power, two controllers were required. These were the remote master controller and the remote accelerator controller. Both could be installed in the turret. If the vehicle was being maneuvered in close quarters where the power requirements did not exceed the engine output at 1300 rpm, only the remote master controller was needed. This unit

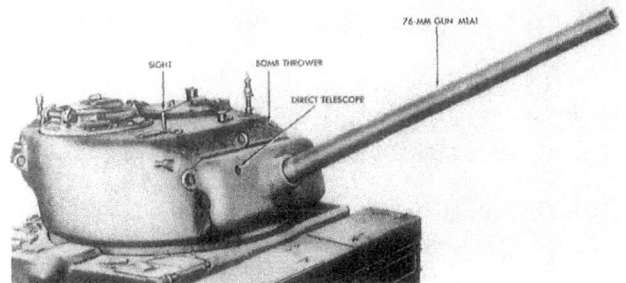

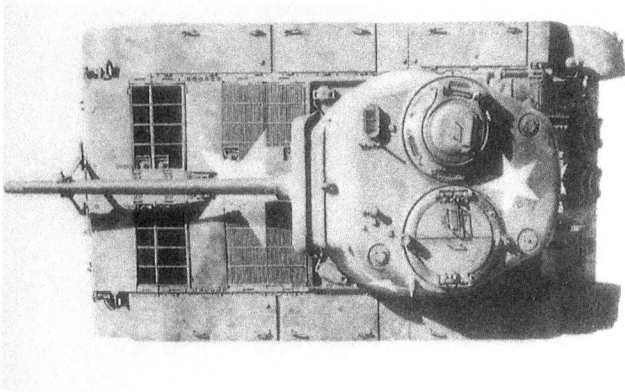

on the end of its cable could easily be carried alongside the tank. The remote control feature did not find favor with the Armored Board who considered it unnecessary and objected to the necessity for locking the turret when driving from that position. It was recommended that the remote control equipment be eliminated and the stowage space be used for more essential items.

Service tests of ten production T23s at Fort Knox concluded that the tank was unsatisfactory for combat use. Difficulties in maintenance together with the retraining necessary for maintenance personnel combined to cause rejection of the vehicle. A similar rejection occurred in February 1945 when the European Theater was informally advised that approximately 200 of the T23s could be made available. In refusing these tanks, it was noted that their introduction would require a complete complement of trained personnel as well as a new inventory of spare parts. Such a disadvantage could not be justified at that time.

The fourth production T23 (serial number 6) during tests at the G.M. Proving Ground. The pistol port has now reappeared, but this tank is not equipped with lugs for the power train hoisting device. The vertical volute spring Sherman suspension was standard for the production tanks. This vehicle is equipped with double pin T48 tracks.

The ninth production T23 (serial number 11) at Aberdeen Proving Ground. The hoisting device lugs are fitted on the right side of the turret.

Production T23 (serial number 55) during desert tests (above and right) near Phoenix, Arizona.

T23 waterproofed and fitted for deep wading tests.

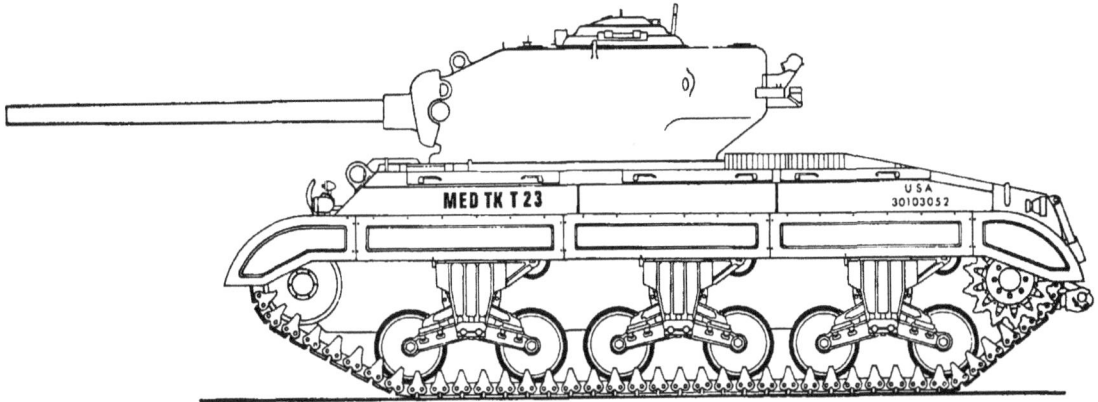

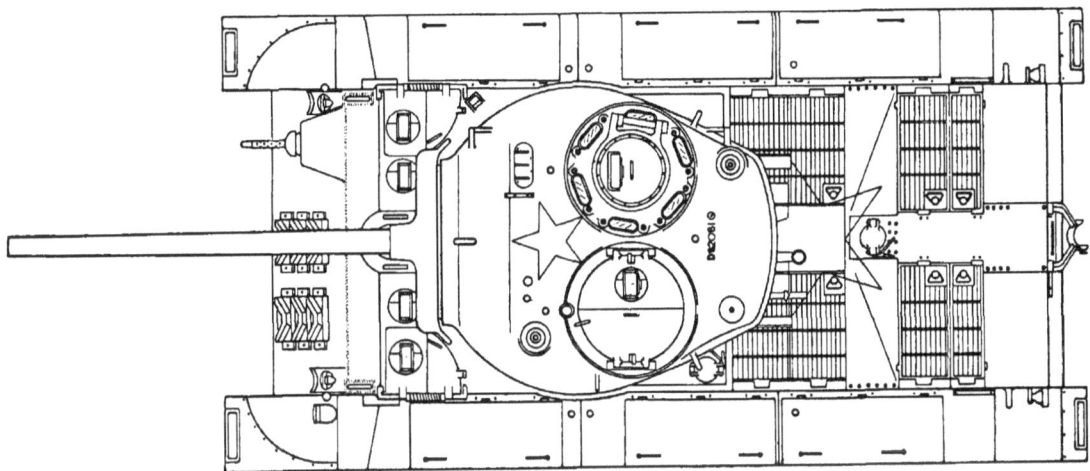

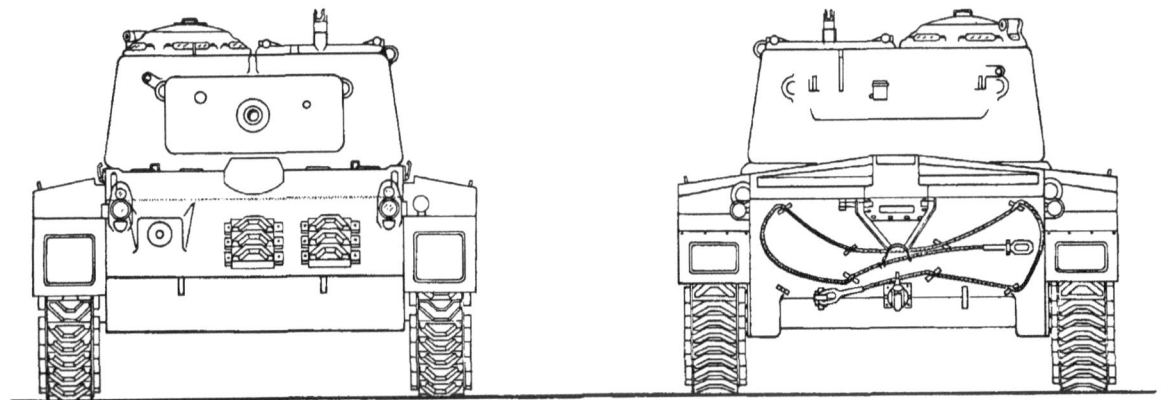

MED TK T23

USA
30103052

Scale 1:48

Medium Tank T23, Production Model

85

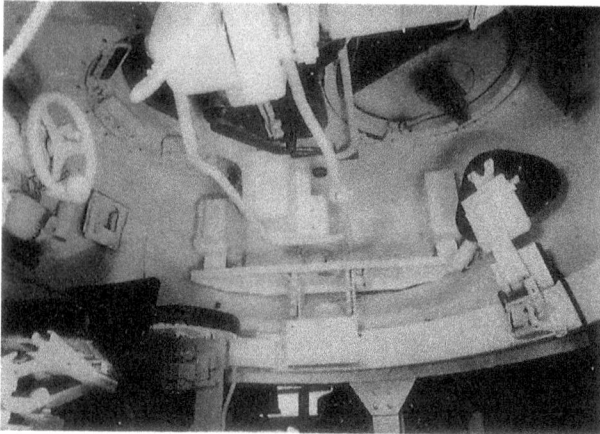

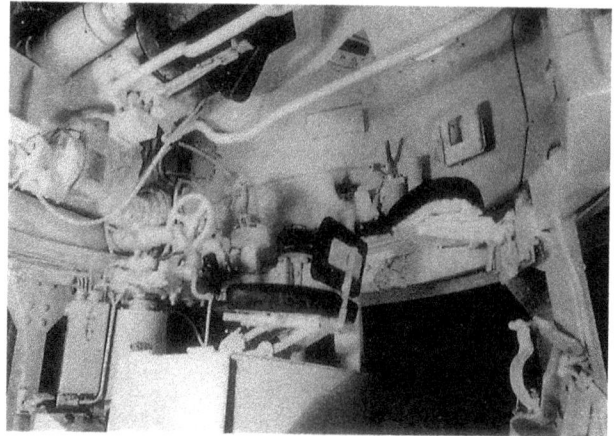

In the view of the turret rear (above left) the radio has not been installed and the rack in the bustle is empty. The right side of the turret (above right) shows the gunner's seat with the tank commander's lower seat immediately in the rear. Other views of turret details are shown below with the 76 mm gun M1A1 and the M62 mount at the bottom.

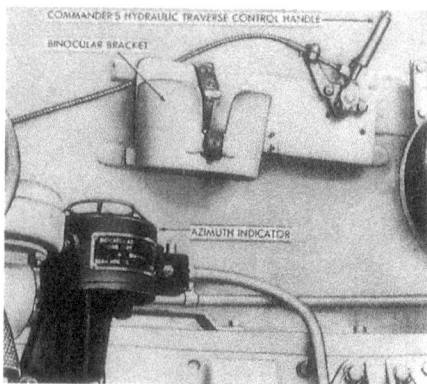

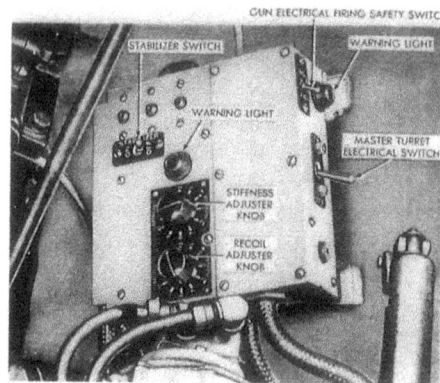

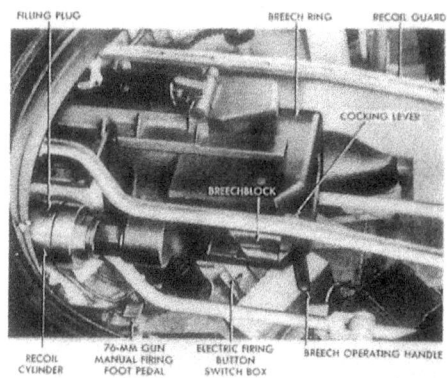

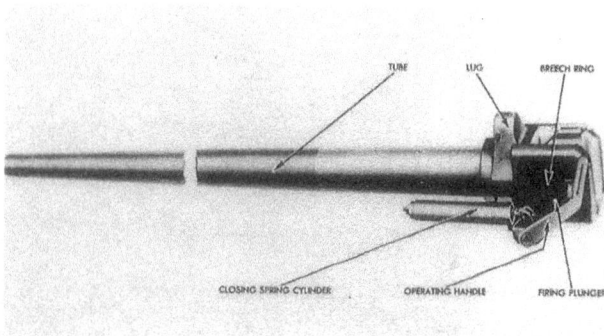

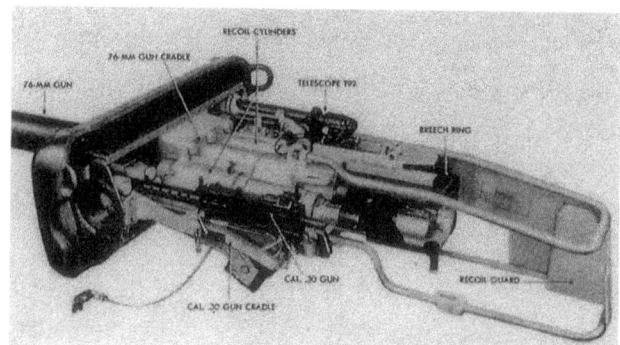

LEFT MASTER CONTROLLERS — SIREN SWITCH — ACCELERATOR PEDAL — PORTABLE FIRE EXTINGUISHER — PRIMER PUMP — ESCAPE DOOR — PARKING BRAKE — SEAT

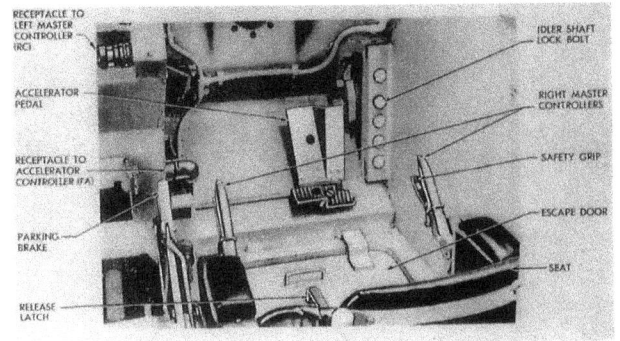

RECEPTACLE TO LEFT MASTER CONTROLLER (RC) — IDLER SHAFT LOCK BOLT — ACCELERATOR PEDAL — RIGHT MASTER CONTROLLERS — RECEPTACLE TO ACCELERATOR CONTROLLER (FA) — SAFETY GRIP — PARKING BRAKE — ESCAPE DOOR — SEAT — RELEASE LATCH

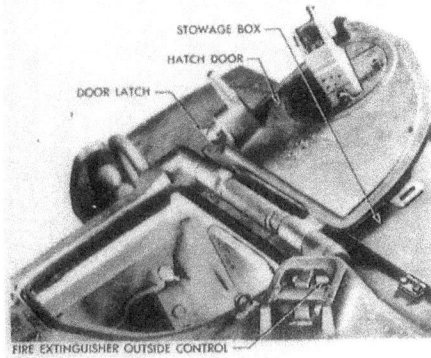

STOWAGE BOX — HATCH DOOR — DOOR LATCH — FIRE EXTINGUISHER OUTSIDE CONTROL

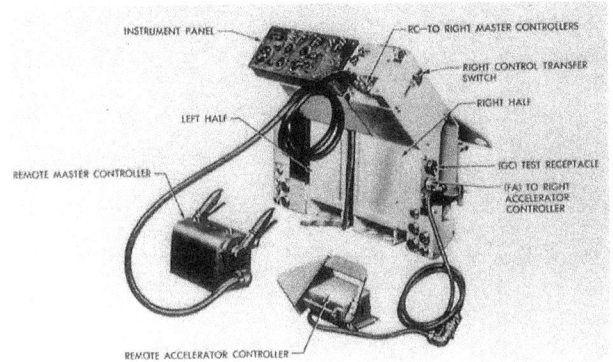

INSTRUMENT PANEL — RC—TO RIGHT MASTER CONTROLLERS — RIGHT CONTROL TRANSFER SWITCH — LEFT HALF — RIGHT HALF — REMOTE MASTER CONTROLLER — (GC) TEST RECEPTACLE — (FA) TO RIGHT ACCELERATOR CONTROLLER — REMOTE ACCELERATOR CONTROLLER

The driver's (top left) and assistant driver's (top right) positions are shown above along with hatch details and views of the instrument panel and remote controllers.

Engine compartment cover arrangement and power unit removal are illustrated immediately below. Note the engine and propulsion generator are removeable as one unit.

The photograph at bottom left shows the final drive assembly, while the one at bottom right demonstrates the method of setting the track tension using the adjustable idler.

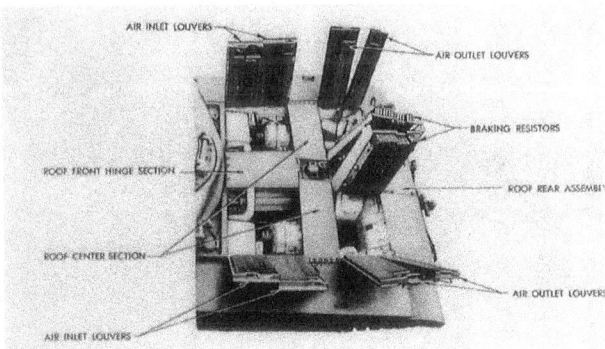

AIR INLET LOUVERS — AIR OUTLET LOUVERS — BRAKING RESISTORS — ROOF FRONT HINGE SECTION — ROOF REAR ASSEMBLY — ROOF CENTER SECTION — AIR OUTLET LOUVERS — AIR INLET LOUVERS

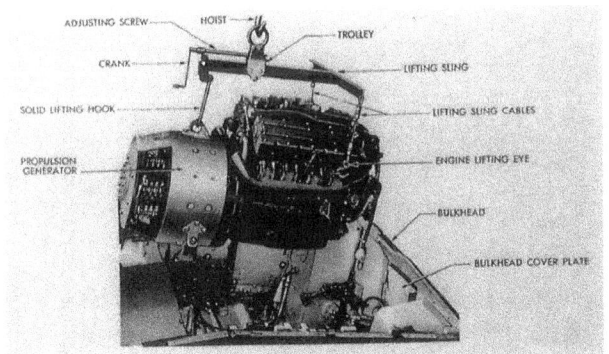

ADJUSTING SCREW — HOIST — TROLLEY — CRANK — LIFTING SLING — SOLID LIFTING HOOK — LIFTING SLING CABLES — PROPULSION GENERATOR — ENGINE LIFTING EYE — BULKHEAD — BULKHEAD COVER PLATE

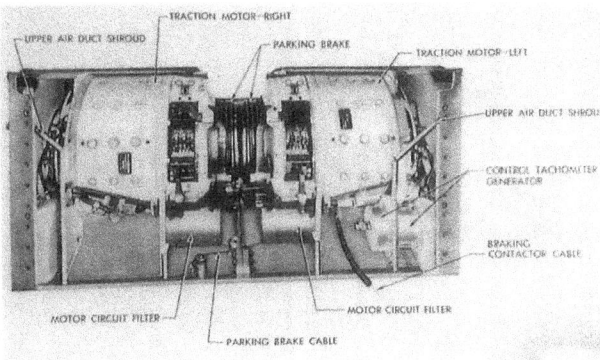

TRACTION MOTOR—RIGHT — UPPER AIR DUCT SHROUD — PARKING BRAKE — TRACTION MOTOR—LEFT — UPPER AIR DUCT SHROUD — CONTROL TACHOMETER GENERATOR — BRAKING CONTACTOR CABLE — MOTOR CIRCUIT FILTER — MOTOR CIRCUIT FILTER — PARKING BRAKE CABLE

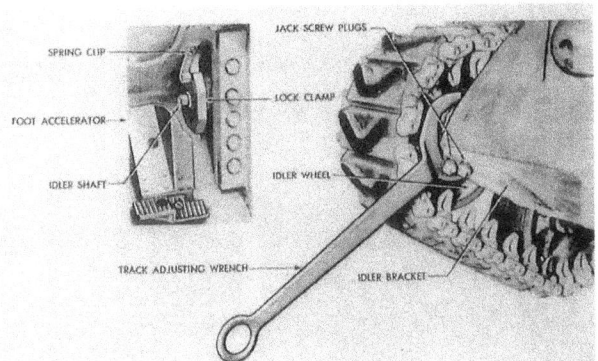

JACK SCREW PLUGS — SPRING CLIP — LOCK CLAMP — FOOT ACCELERATOR — IDLER SHAFT — IDLER WHEEL — TRACK ADJUSTING WRENCH — IDLER BRACKET

87

The pilot T23E3 under test by the Armored Board at Fort Knox, Kentucky.

In April 1943 the Ordnance Committee recommended that pilot models of the T23 be constructed using the torsion bar suspension and it designated such pilots as medium tanks T23E3. This action was approved in December 1943 and the Chrysler Corporation was authorized to construct two pilot tanks. A change later limited the construction to only one vehicle. The pilot medium tank T23E3 (registration number 30103068) was completed and delivered to the Detroit Tank Arsenal on August 29, 1944 for shipment to Fort Knox.

The torsion bar suspension ran on 19 inch wide center guided tracks and was equipped with six dual road wheels on each side. The first two and last two road wheels on each side were fitted with shock absorbers and five dual track return rollers were provided for each track.

The turret and much of the equipment used in the construction of the T23E3 were obtained from production T23 number 19, whose registration number it also acquired. This vehicle had been manufactured at the Detroit Tank Arsenal and was diverted to the Tank Laboratory for disassembly. T23 number 19

was manufactured before the water proofing specifications were released, however, all the electrical equipment was water-proofed before being installed in the T23E3. The turret was also brought up to date by removing the basket and relocating the equipment previously supported by the turret platform. The track and suspension used on the new pilot were obtained from a T25E1 medium tank at Aberdeen Proving Ground. A new hull was manufactured for the T23E3 and the original hull from T23 number 19 was shipped to Aberdeen Proving Ground for ballistic tests. After completion, the tank was shipped to the Armored Board at Fort Knox. However, no production resulted due to lack of interest in electric drive tanks.

In July 1943, it had been proposed to standardize the T23E3 as the medium tank M27 with the electric drive and torsion bar suspension. At the same time it was proposed to standardize the T20E3 as the medium tank M27B1 with the torqmatic drive and torsion bar suspension. An OCM item was drafted, but standardization was not approved.

The T23E3 pilot after completion at Chrysler in August 1944.

The photographs above and to the right show the pilot T23E3 shortly after completion at Chrysler. Unlike the earlier T20E3, this tank has a 13 tooth sprocket. Stowage was also modernized to the latest practice, including the addition of a camouflage net rack on the right side of the turret. The box on the left side of the turret, just forward of the pistol port, was for stowage of a first aid kit.

At the left, the T23E3 is fitted with a double pin center guided track. Note that the sprocket has also been modified.

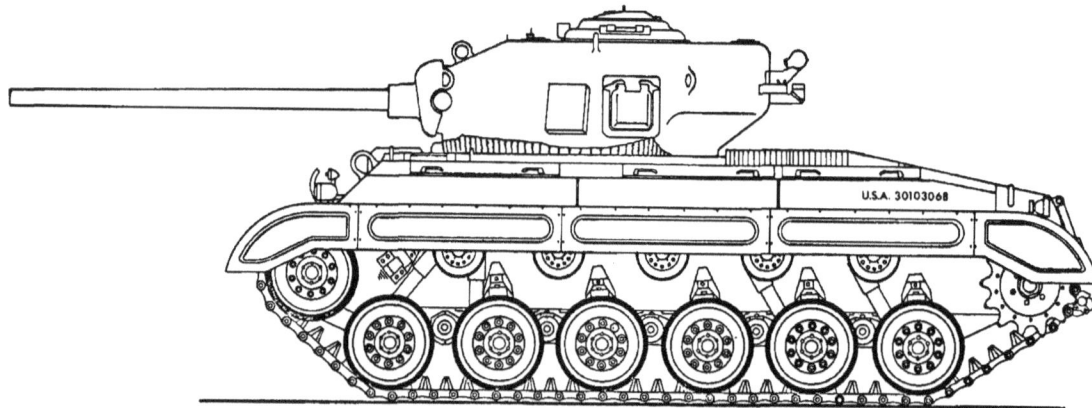

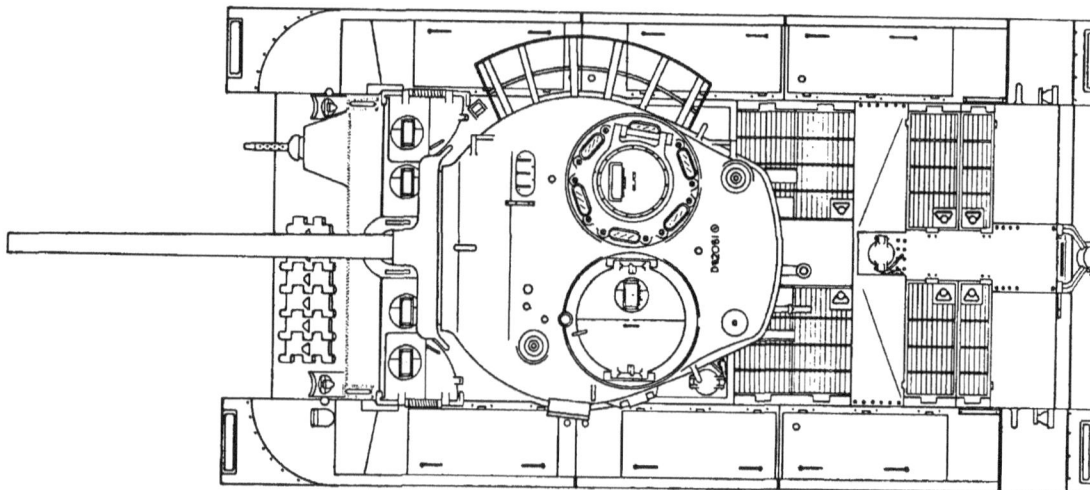

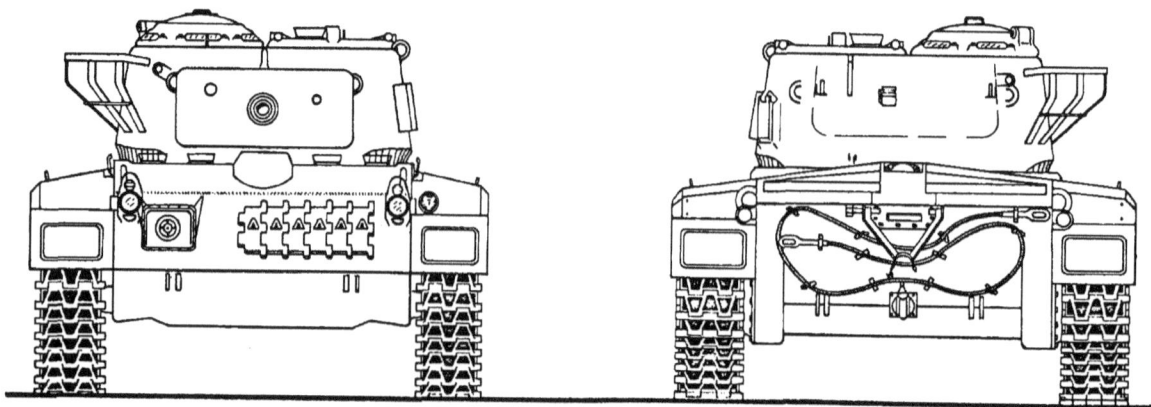

U.S.A. 30103068

Scale 1:48

Medium Tank T23E3, Pilot

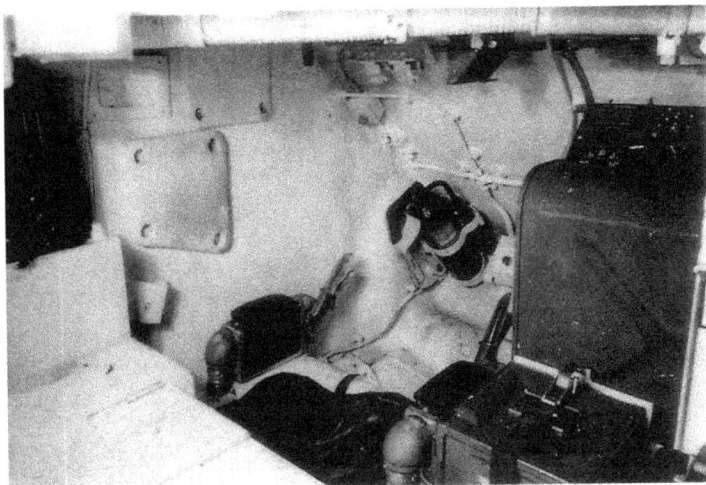

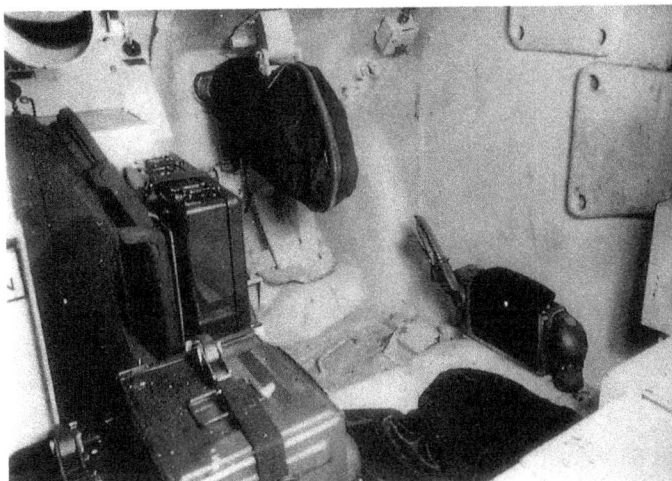

The driver (top left) and assistant driver (top right) were equipped with same controls as in the production T23. A canvas cover has been placed over the .30 caliber bow machine gun. The turret stowage (center) was modified to conform to the design practice of August 1944 with the ready rounds held in a rack suspended from the turret ring. The turret basket was removed and the gunner's seat (bottom) suspended from the turret ring. The ammunition was stowed in floor racks made accessible by the elimination of the turret basket.

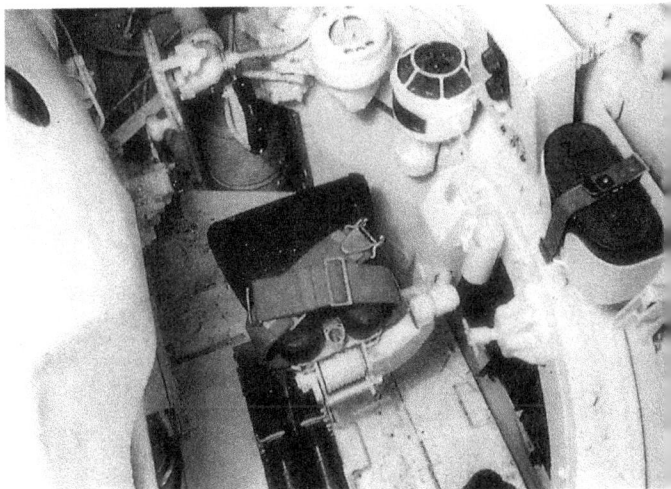

A further variation in suspension systems was proposed for the T23 with the introduction of the pilot medium tanks M4E8. These vehicles were equipped with horizontal volute spring suspensions and 23 inch center guided tracks. The Ordnance Committee recommended that a pilot medium tank T23 be built with this suspension system and designated such a pilot as the medium tank T23E4. The Corps of Engineers objected since the vehicle would be 131 inches wide, exceeding by seven inches the permissible width under AR850-15. Because of this excess width, the procurement of the pilot was disapproved. However, at a later date, three production T23s were fitted with the horizontal volute spring suspension and 23 inch T80 tracks and shipped to Fort Knox for test purposes. The tests reported, not surprisingly, that the wide track horizontal volute spring suspension was superior in performance to the standard 16-9/16 inch track and vertical volute spring suspension as provided on the standard medium tank T23. Again, no production resulted due to the end of the war and lack of interest in the electric drive vehicles.

The four views at the left show one (registration number 30112588) of the three production T23s modified to use the horizontal volute spring suspension and center guided track. The tanks were under test by the Armored Board and were fitted with the T80 double pin track. The guns on these tanks were equipped with muzzle brakes and a later type of traveling lock.

Another of the three tanks (registration number 30112590) is shown below in its final resting place at Aberdeen Proving Ground.

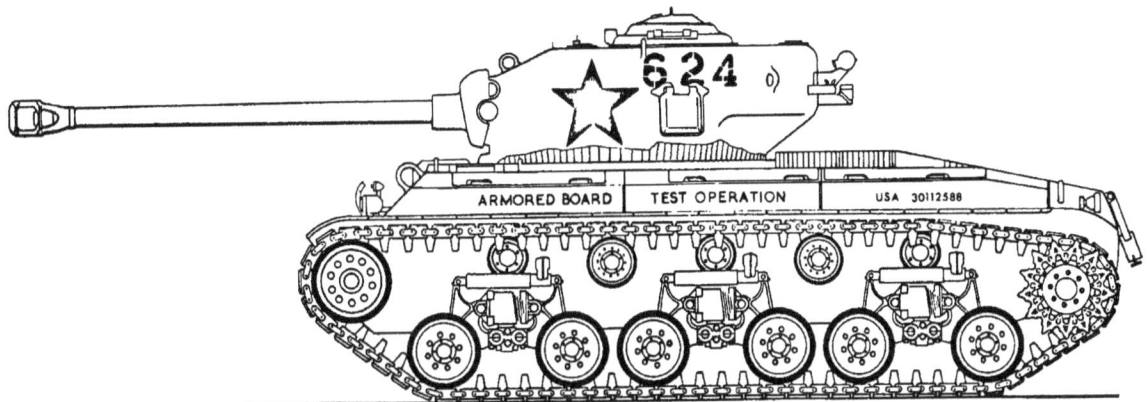

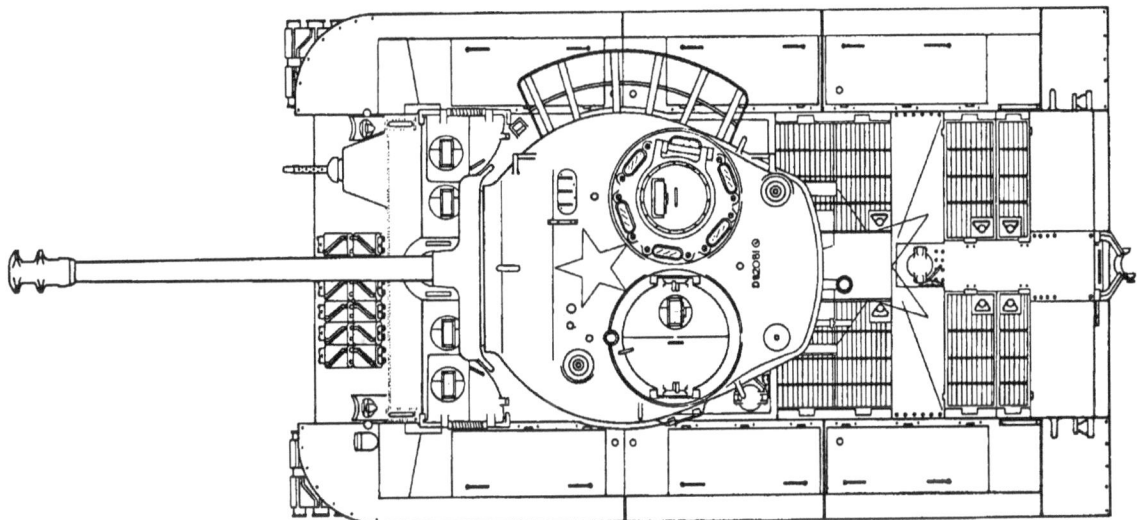

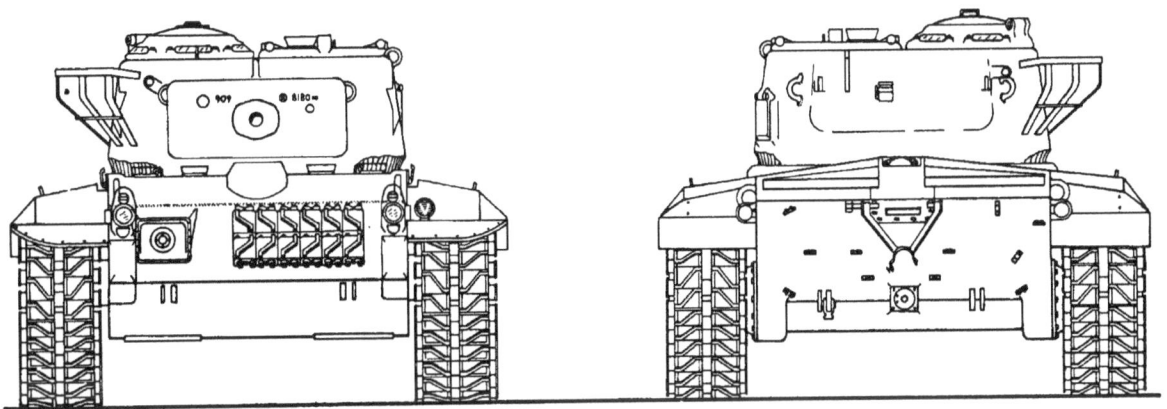

ARMORED BOARD | TEST OPERATION | USA 30112588

Scale 1:48

Medium Tank T23 HVSS

T25 pilot number 1 at Chrysler in January 1944. The same tank is shown below with the hatches open.

THE T25 AND T26

At the time production was authorized for the T23 medium tank, 50 of these vehicles were requested modified to mount the 90 mm gun. Forty tanks, using the same armor basis as the T23, were designated medium tank T25. The remaining ten vehicles were to be equipped with heavier armor and designated medium tank T26. The latter was expected to equal the firepower and protection of the German Panzerkampfwagen VI Tiger I and still maintain a lower vehicle weight. The original design estimates indicated a weight of 72,000 pounds for the T25 and 80,000 pounds for the T26. However, continued development showed an increase in weight for the T25 to about 81,000 pounds and a similar large increase for the T26. Since a great portion of the weight was due to the electric transmission, a decision was made to replace the electric drive with a torqmatic transmission. The modified designs were designated medium tanks T25E1 and T26E1 and they were selected to fill the production order for 50 tanks in place of the previous models.

Following the decision to change the limited procurement order to the T25E1 and T26E1, it was requested that two T23s be converted to the original design concept for the T25, mounting a 90 mm gun turret. Medium tank T25 pilot number 1 (registration number 30103053) was completed by the Chrysler Engineering Division and delivered to Detroit Tank Arsenal for shipment to Aberdeen Proving Ground on January 21, 1944. T25 pilot number 2 (registration number 30103054) was completed and delivered to the Tank Arsenal for shipment to Fort Knox on April 29, 1944. Pilots 1 and 2 weighed 82,310 lbs. and 84,210 lbs. respectively. Both vehicles were equipped with the horizontal volute spring suspension and the

23 inch center guided track. The major portion of the weight difference was due to a weight increase in the suspension and tracks as a result of redesigning the main brackets, increased tire thickness, and changing the track sections. The balance of the difference may be attributed to numerous minor changes and variations in casting thickness.

The T25s carried a 90 mm gun T7 and a coaxial .30 cal. machine gun mounted in a larger, T25E1 type turret. The 90 was not equipped with a stabilizer in either the T25 or T26. The hull differed from that of the T23 in having longer and narrower drivers' hatch openings to provide clearance past the gun shield when opening the doors with the gun forward. It also had a stronger ribbed roof casting to support the heavier turret assembly. The 3 inch thick hull front was a welded assembly of castings and rolled plate identical in appearance to the production model T23. Rear hull braces were eliminated to provide more ammunition stowage space. The power plant and propulsion equipment consisted of the Ford V-8 Model GAN tank engine with electric drive, identical with the T23. Testing of the medium tank T25 began at the Armored Board on September 28, 1944. By this time, interest was centered on the heavier T26E1 vehicle and at the conclusion of the tests, it was recommended that no further consideration be given to medium tank T25.

T25 pilot number 1 under test at Aberdeen Proving Ground. Pilot number 1 had no pistol port and was not fitted with a net stowage rack on the right side of the turret. The gun traveling lock was the open top type, similar to that on the T23 early production tanks. The horizontal volute spring suspension with the 23 inch T66, center guided, single pin track can be seen clearly in these views. Note that shock absorbers are mounted on all three bogies. Only the two large track return rollers are visible in the photographs. In addition, there were three small return rollers per track, one directly above each bogie.

The turret arrangement shown below was a larger, heavier version of the T23, retaining the vision cupola for the tank commander and the rotating hatch for the loader. The 90 mm gun was not equipped with a muzzle brake on either T25 pilot.

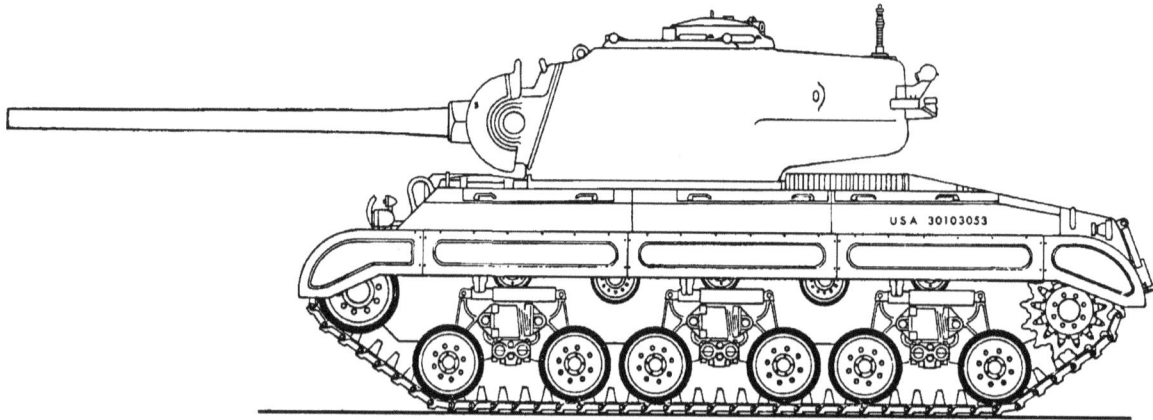

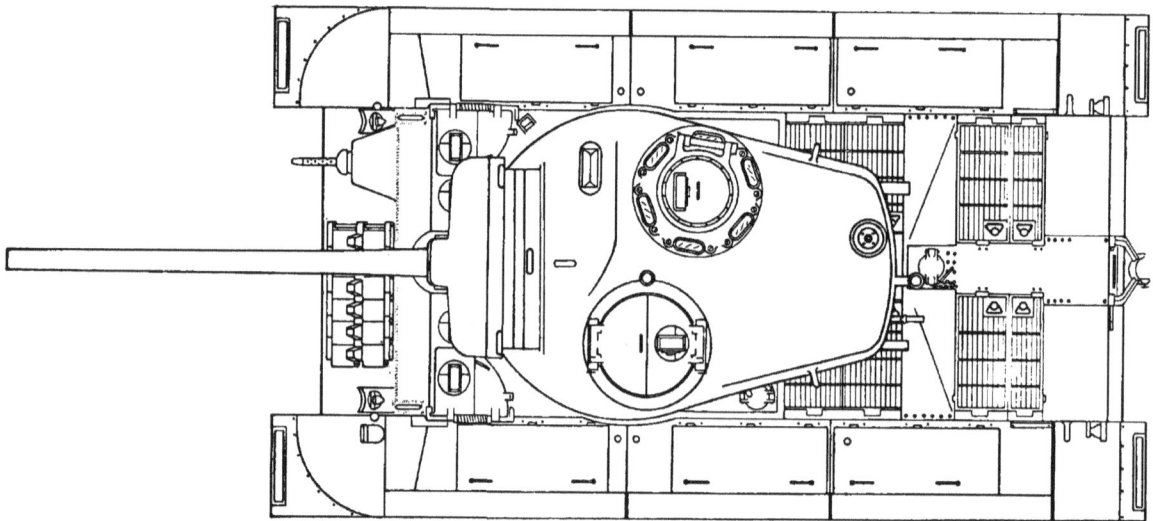

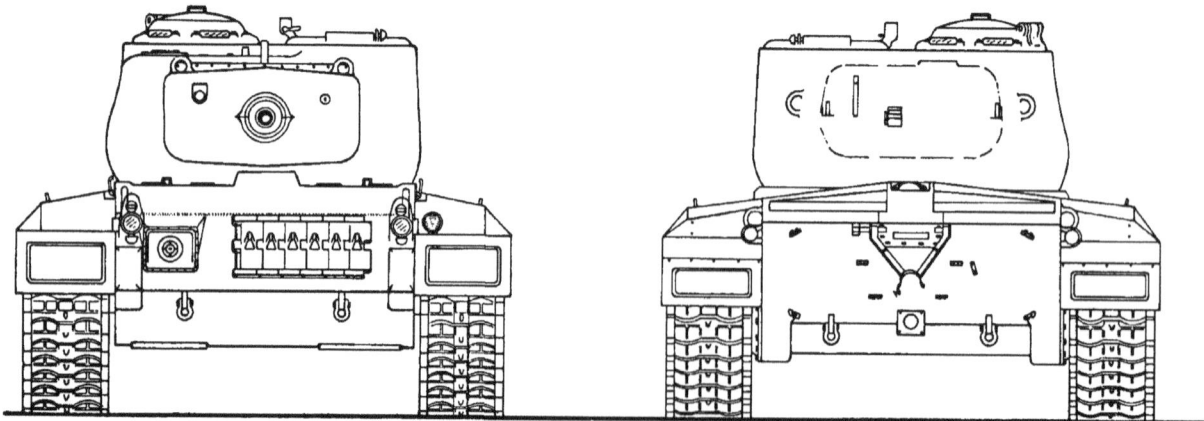

USA 30103053

Scale 1:48

Medium Tank T25, Pilot Number 1

T25 pilot number 2 at Chrysler in April 1944. The tank is readily identified by the turret net stowage rack and pistol port. A closeup at the left shows the traveling gun lock similar to that on the late production T23. Note the weld metal arrows on the rear hull to indicate proper stowage of the towing cable. The words START and FINISH are partially obscured by the cable itself.

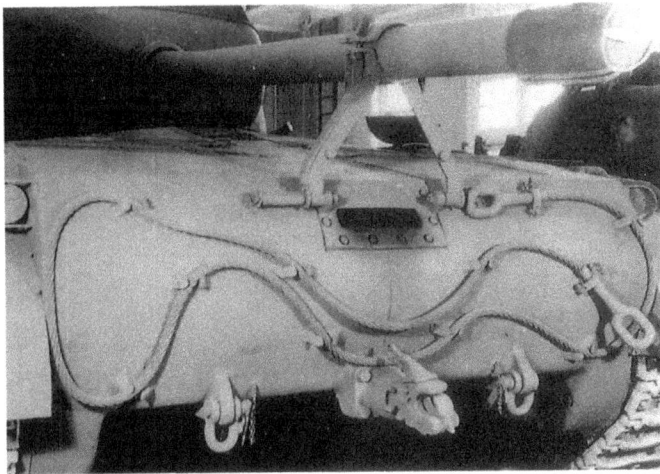

Pilot number 2 under test by the Armored Board at Fort Knox. As happened frequently during these tests, the sandshields have been removed. The tow cable has also been lost along the way and the tank is now equipped with T80 double pin tracks. However, six T66 single pin track blocks are still carried on the front armor plate.

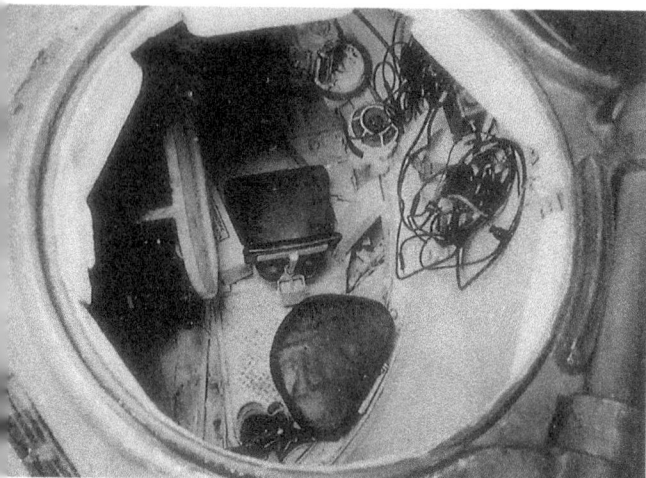

The view through the tank commander's hatch is compared above for T25s pilot number 1 (left) and pilot number 2 (right). The partial turret basket was removed on pilot number 2 and the floor ammunition stowage rearranged. The gunner's seat was suspended from the turret ring on the later vehicle. The recoil guard and breech of the 90 mm gun is visible at the left of each photograph.

The loader's station is compared for pilot number 1 (below) and pilot number 2 (right). On pilot number 2 the 90 mm rounds were enclosed in canvas containers. One 90 mm high explosive shell is shown in the ready rack of pilot number 1.

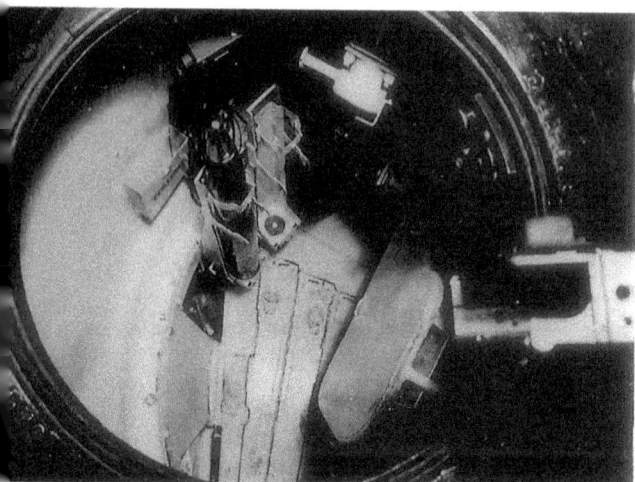

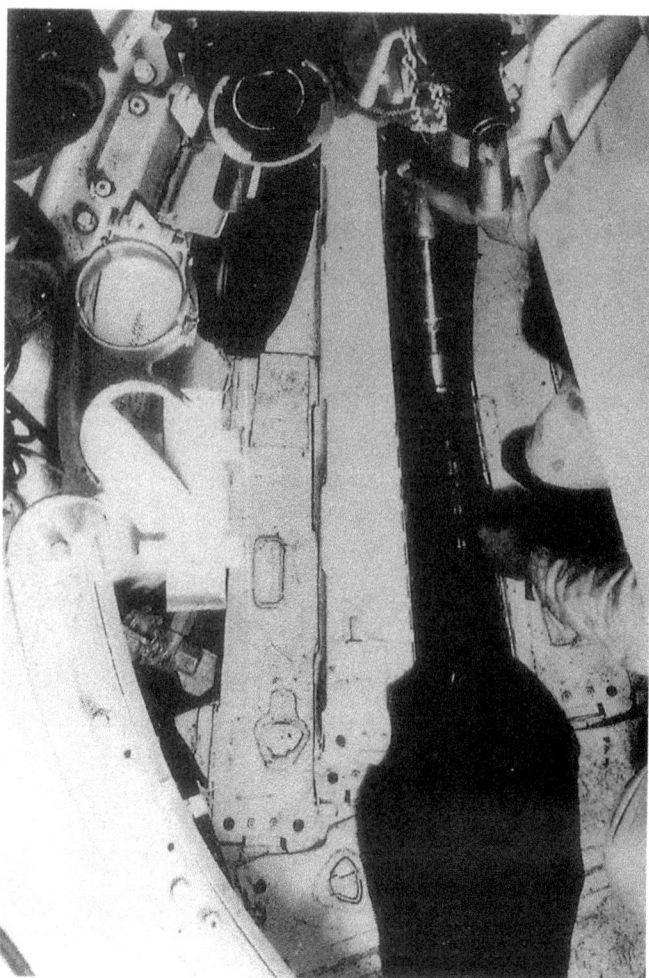

The driver's (left) and assistant driver's positions are shown above for pilot number 2. Note the similarity to the late production T23. The controls were identical.

Further details of the turret stowage arrangements on pilot number 2 are illustrated below. At the loader's station (left) the .30 caliber coaxial machine gun is clearly visible with its ammunition box. The 2 inch smoke mortar is in the open position between the machine gun and the 90 mm ready round. The radio and other bustle stowage are shown immediately below (right) followed by the gunner's station (bottom right).

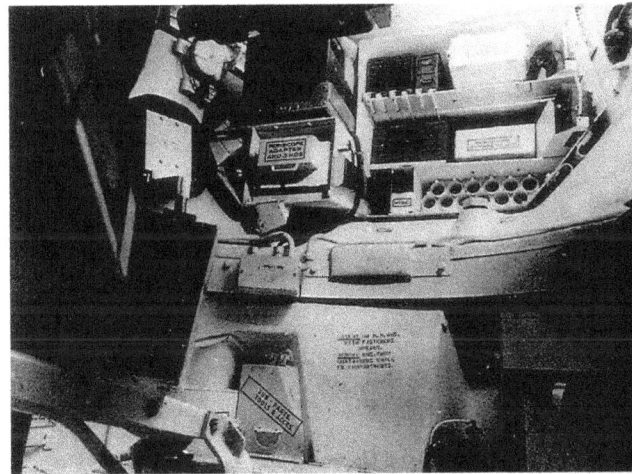

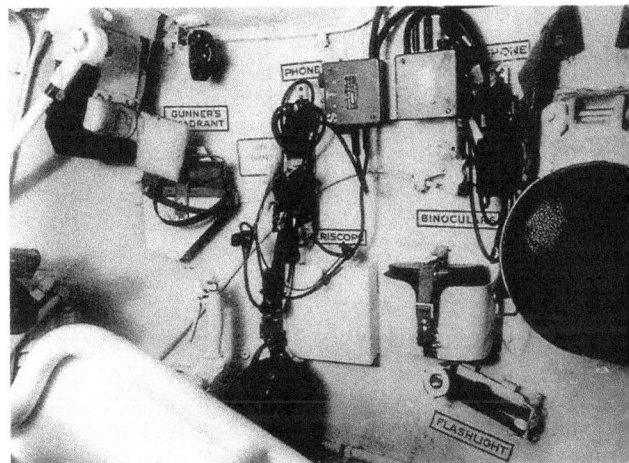

Although the T26 was eliminated in favor of the T26E1 for the limited procurement order of ten tanks, the Office, Chief of Ordnance, authorized the procurement of pilots according to the original T26 concept for test purposes. These vehicles were to be equipped with the electric drive and a torsion bar suspension.

The original directive from the Ordnance Department authorized the building of two T26 pilots. However, the construction program was reduced to one on May 6, 1944. This tank, (registration number 30128307) was delivered by the Tank Laboratory of the Chrysler Engineering Division for shipment to Fort Knox on October 28, 1944. Its combat weight was 95,100 pounds. The torsion bar suspension and the 24 inch track were salvaged from a T26E1 at Aberdeen Proving Ground and a few modifications were made prior to its installation. The power plant and electric drive were identical to the system in the medium tank T23. Front hull construction differed somewhat from the T25, consisting of a welded assembly of rolled plates similar in appearance to the T23 pilot models. The front hull armor was 4 inches thick at angle of 46 degrees from the vertical. The T26E1 type turret mounting the 90 mm gun was modified to remove the basket and to make other changes in accordance with the latest turret design.

The T26 was placed under test at Fort Knox, starting on November 30, 1944, in comparison with the T26E1 and the later model T26E3. The T26 was found to be superior in normal terrain operation. However, the T26E1 and T26E3 were superior in reliability, ease of control on steep winding grades, and had smaller maintenance requirements. The Armored Board concluded that the fast forward and reverse speeds, quick acceleration, and

maneuverability of the T26 over normal terrain were superior to that of the T26E1 or the T26E3, but the electric drive was too complicated for the average mechanic to understand or repair. This reflected the general attitude at that time toward the electric drive tanks. As a result of these conclusions, no further work was done on the T26.

T26 pilot under test at Fort Knox.

The T26 turret was similar in design to the T25, with thicker armor, and was fitted with a pistol port. The same arrangement of a vision cupola for the tank commander and the rotating hatch for the loader was retained. The camouflage net bracket appears to be overloaded in these views. This may explain the Armored Board's objection to the location of the rack. Their complaint was that the rack obscured the view from the vision cupola and this was certainly true if it was poorly stowed as shown in the top photograph. The 90 mm gun did not have a muzzle brake on this vehicle. The T26E1 torsion bar suspension with the 24 inch T81 single pin tracks is clearly visible in these photographs. Note the extra T81 track blocks fixed to the front armor below at the right. This view also shows the front hull construction which was welded up from sections of rolled plate. Once again the sandshields have been partially removed during the tests at Fort Knox.

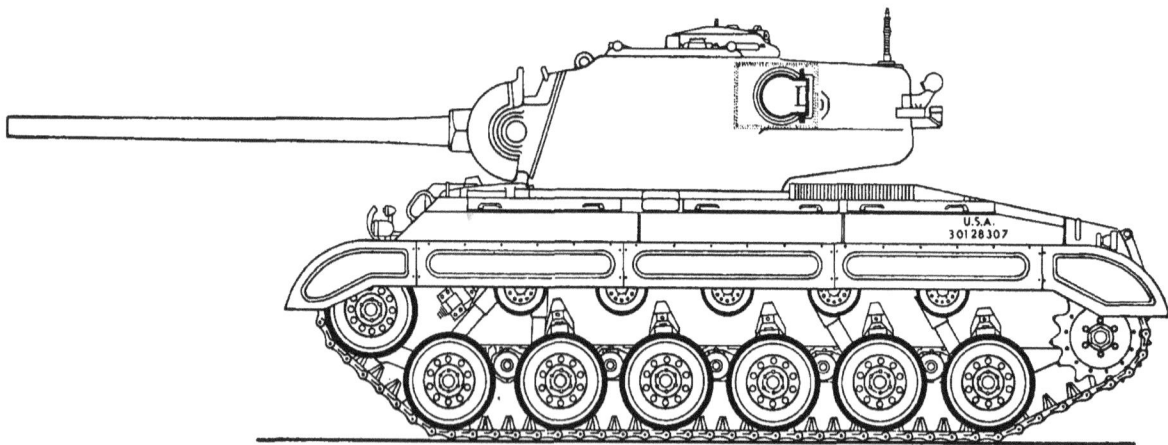

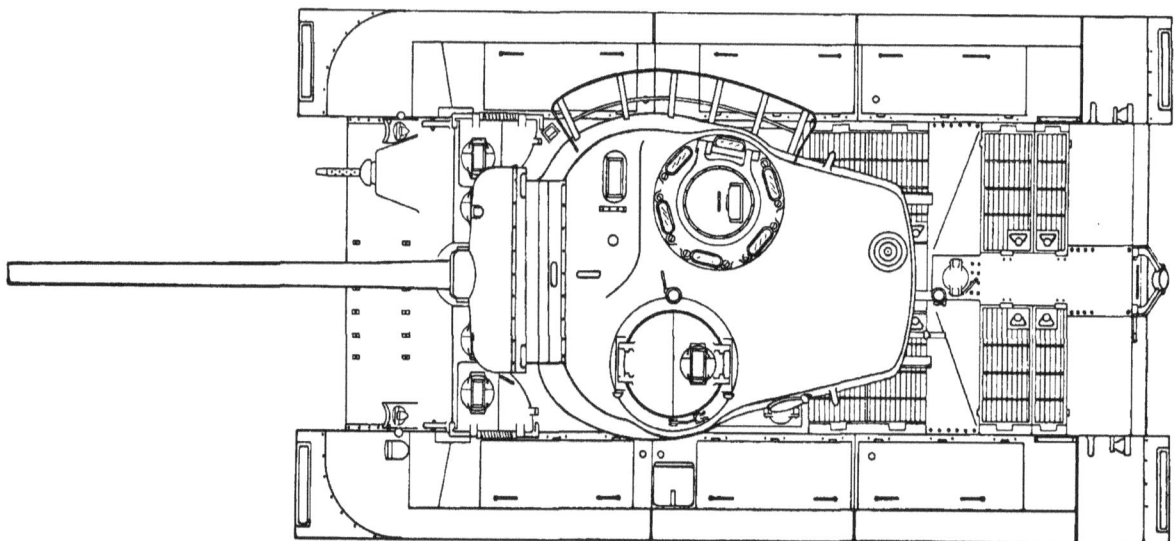

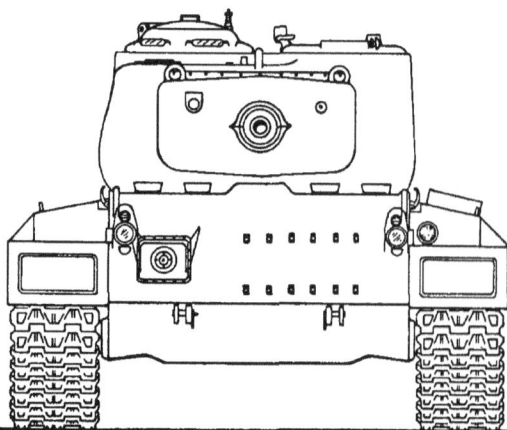

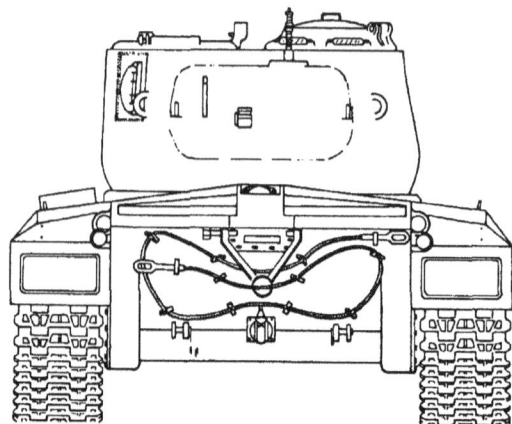

Scale 1:48

Medium Tank T26, Pilot

Details of the outside stowage boxes can be seen in the two views at the right. The upper photograph also shows the pistol port in the open position.

The rear hull and the gun traveling lock are shown below at the right. The opening below the gun travel lock support is the engine exhaust. Like the T25, the tow cable was stowed on the rear hull.

Directly below is a view through the loader's hatch showing the ten round ready rack for the 90 mm gun. The coaxial .30 caliber machine gun appears at the upper right in the photograph. Note there is no turret basket in this vehicle and the covers of the floor ammunition stowage boxes can be seen just below the ready rack.

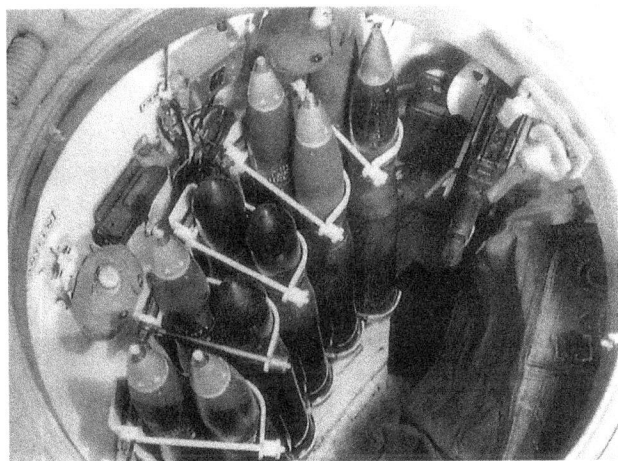

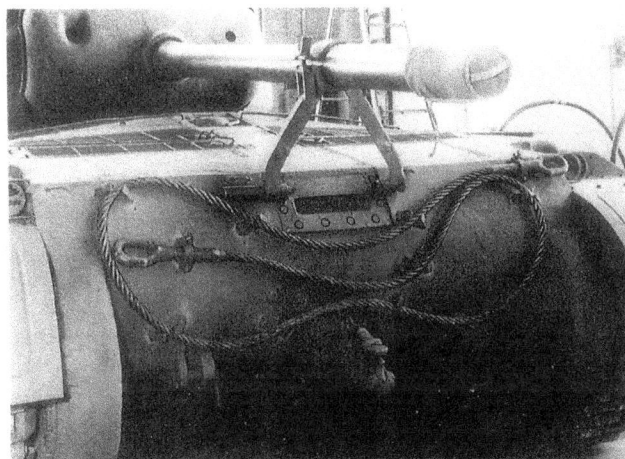

T25E1 number 1 (above and below) at Aberdeen Proving Ground on March 8, 1944.

THE T25E1, T26E1, AND M26 (T26E3)

Deliveries under the limited procurement order for 40 T25E1s and ten T26E1s began in February and ran through May 1944. The first tank, a T25E1, was demonstrated on January 13. All 50 vehicles were built at General Motors Fisher Tank Arsenal, Grand Blanc, Michigan. Of the first T25E1s off the production line, three were shipped to Aberdeen Proving Ground, two to the General Motors Proving Ground, one to Phoenix, Arizona, and five to the Armored Board at Fort Knox. The first T26E1 was forwarded to Aberdeen Proving Ground in February followed shortly by number 2.

At a later date, T26E1 number 5 was shipped to Fort Knox for evaluation by the Armored Board. The registration numbers ranged from 30103252 through 30103291 for the 40 T25E1s and from 30103292 through 30103301 for the ten T26E1s. The two models were quite similar in appearance and construction, both being equipped with the Ford GAF tank engine and torqmatic transmission. The 70,000 lb. unstowed weight of the T25E1 rose to 81,300 lbs. for the T26E1, due to the latter's heavier armor. This in turn required an increase in track width from 19 to 24 inches and a change in the final drive gear ratio.

The hull was an assembly of armor steel castings or sections welded to rolled armor plate. A single large V-shaped casting was used for the front hull in both tanks, differing only in thickness. The upper front was 3 inches thick on the T25E1 and 4 inches thick on the T26E1, both at 46 degrees from the vertical. The V-shaped front casting was welded to

the floor at the bottom and extended up and back past the drivers' hatches to form the front of the turret ring opening. The bulge for the .30 caliber bow machine gun mount was an integral part of the casting. The rear of the turret ring opening was formed by the rear upper hull casting which extended across the width of the tank. Both castings were welded to the center upper hull plates and the hull side plates. The cast rear hull side plates, containing a large machined hole for the final drives, were welded between the hull side plates and the hull rear plate. The floor side plates were welded to the hull floor at an angle slanting upward to meet the hull side plates giving a boat shaped bottom to the hull. The wheel arm supports extended through openings in the floor side plates and were bolted to the outside. Crossmembers were welded to the inside hull floor for reinforcement and to support equipment above the torsion bars. On top of the hull, a cast transverse

104

All three photographs on this page show T26E1 number 1 under test at Aberdeen Proving Ground. This was the tank later modified to mount the 90 mm T15E1 gun and then shipped to Europe.

housing extended across the center of the engine compartment and was bolted to the side plates to strengthen the structure. A curved bulkhead was welded in place to separate the fighting compartment from the engine compartment and to support the hull top plate and rear section of the turret ring. Two escape hatches, one in front of each driver's seat, provided emergency exits.

Dual controls for the driver and assistant driver were mounted in the left and right front hull respectively. Since the controlled differential was located at the rear of the engine compartment, the steering levers were connected by a series of linkage rods extending to the rear between the ammunition stowage boxes. A .30 caliber M1919A4 machine gun in the bow hull mount was operated by the assistant driver. No sights were fitted on this weapon and tracers were used for aiming, with the fire being observed through a periscope. Four M6 periscopes were installed, two for the driver and two for the assistant driver. Each man had one periscope in his hatch cover and one in the hull roof just inboard of his hatch. Adjustable seats permitted the drivers to ride with their heads out when the hatches were open or, in the lowered position, to observe through the periscopes with the hatches closed. Ventilation for both the driving and fighting compartments was provided by the 400 cfm rotoclone blower located on the hull roof between the drivers. The blower and its intake caused the large bulge in the top center of the front armor casting.

The gun travel lock pivoted on the exhaust casting is clearly visible in the photograph above. Also, note the different tow cable stowage compared to the T25 and T26.

Details of the massive T26E1 turret can be seen above. The .50 caliber machine gun on the loader's hatch is locked in its travel position with the muzzle held by the retaining clip.

105

T25E1 number 4 under test at the General Motors Proving Ground. Without sandshields, the suspension details are clearly visible. Note the shock absorbers on the first two and last two road wheels. The photograph at the left gives a good view of the sprocket showing how it engaged the lugs at the end of each track link. These views also show the 90 mm gun locked in its travel position.

The photograph above and the two views at the left show three T25E1s, numbers 1, 6, and 8, during test operations at Aberdeen Proving Ground. The suspension diassembly above reveals considerable detail including the volute bumper springs used to limit the travel of each road wheel arm. Note the heavy wear on the sprocket teeth.

The view below was taken at Fort Knox, Kentucky during tests by the Armored Board. This is T25E1 number 21. The 601 on the turret was the Armored Board number assigned to the test vehicle.

T26E1 number 4 on test at the General Motors Proving Ground. The number 8537 was assigned to the vehicle by the Proving Ground. Similar four digit numbers can be seen on other vehicles tested at General Motors. The T81 single pin tracks are shown clearly in the two upper views. At the left, the sprocket can be seen engaging the holes near the end of each track link. Compare this with the similar view of the T25E1.

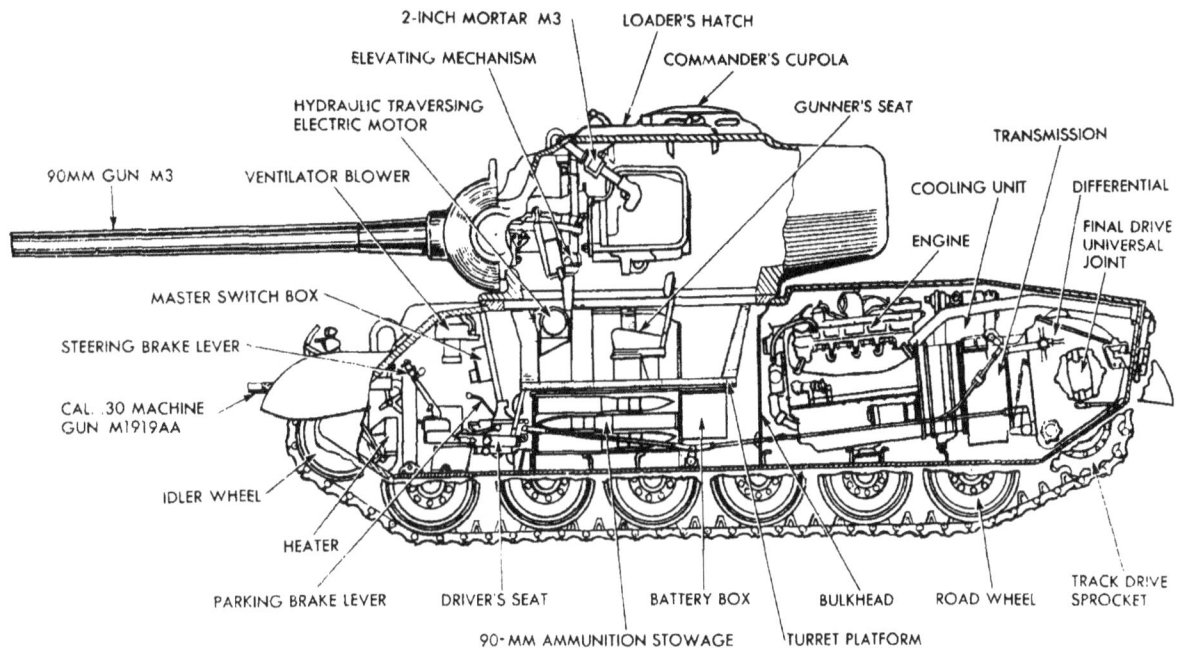

2-INCH MORTAR M3
ELEVATING MECHANISM
HYDRAULIC TRAVERSING ELECTRIC MOTOR
90MM GUN M3
VENTILATOR BLOWER
LOADER'S HATCH
COMMANDER'S CUPOLA
GUNNER'S SEAT
TRANSMISSION
COOLING UNIT
DIFFERENTIAL
FINAL DRIVE UNIVERSAL JOINT
ENGINE
MASTER SWITCH BOX
STEERING BRAKE LEVER
CAL. .30 MACHINE GUN M1919AA
IDLER WHEEL
HEATER
PARKING BRAKE LEVER
DRIVER'S SEAT
BATTERY BOX
BULKHEAD
ROAD WHEEL
TRACK DRIVE SPROCKET
90-MM AMMUNITION STOWAGE
TURRET PLATFORM

Sectional view showing stowage in medium tanks T25E1 and T26E1. Note that all 90 mm ammunition is stowed below the fighting compartment floor.

The photographs at the left show a race over a 4.3 mile course in the Churchville cross country area at Aberdeen Proving Ground between medium tanks T26E1, T25E1, M4A3E8, and the M4A3 with the standard vertical volute spring suspension. The top view shows T25E1 number 6 passing T26E1 number 2.

The T25E1 had the best recorded time of 23 minutes followed by the T26E1 with 26 minutes. The M4A3E8 and M4A3 finished in 28 minutes 35 seconds and 30 minutes 40 seconds respectively. A T23 was also in the race, but threw a track and did not finish.

The Pershing was a tight fit when it came to crossing bridges. This is one method (at right) of crossing a 60 ton Bailey bridge. The heavy timbers were used to protect the bridge curbs.

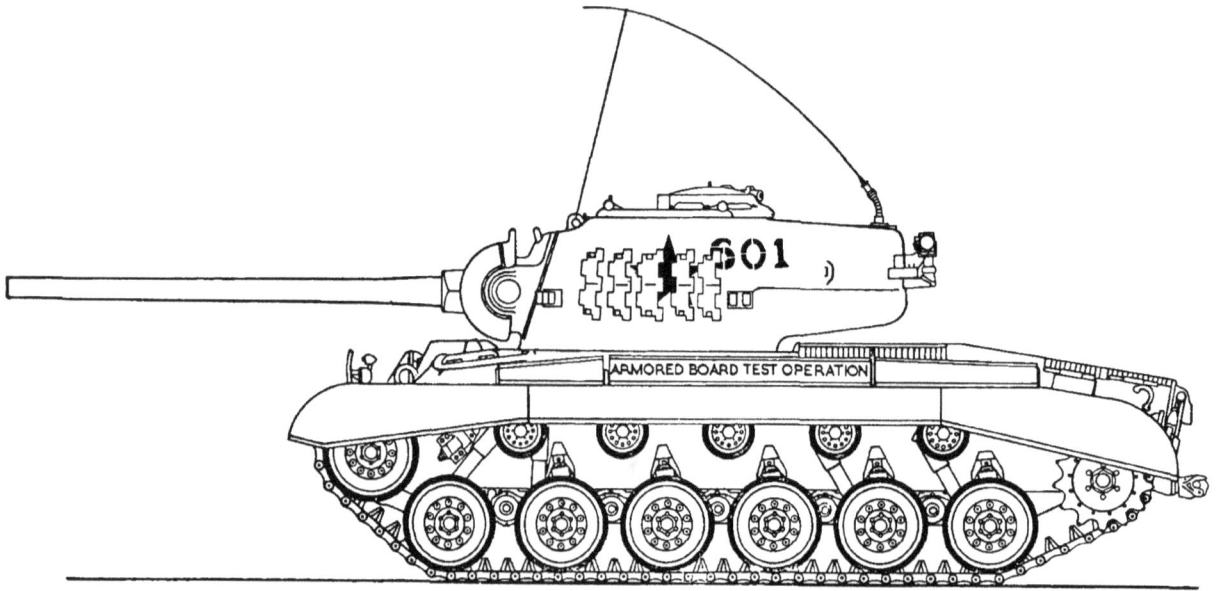

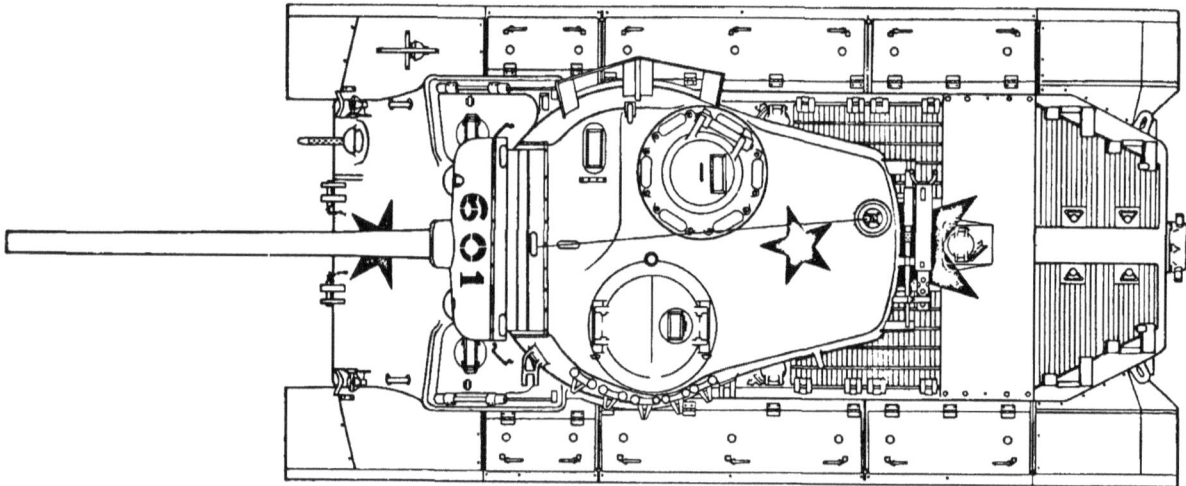

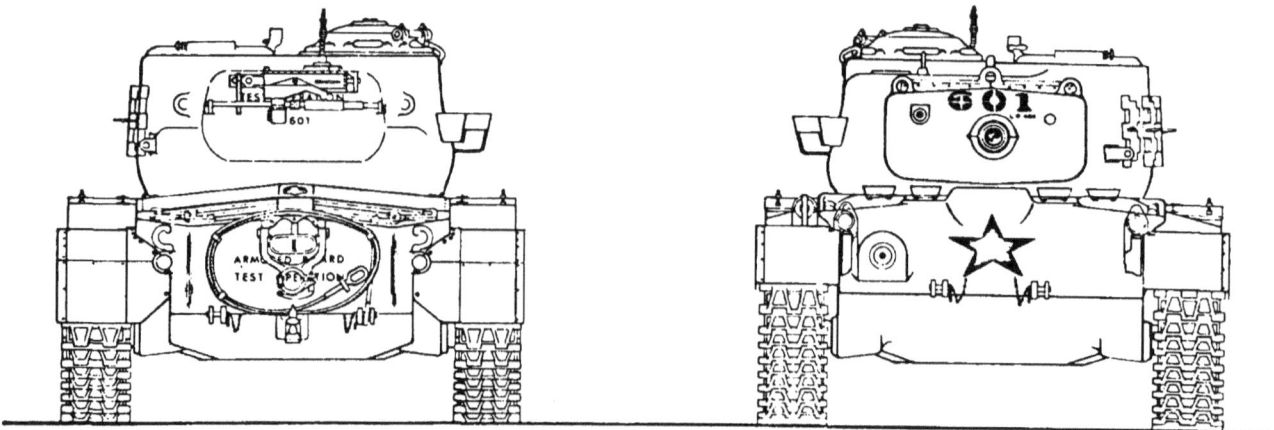

ARMORED BOARD TEST OPERATION

601

Scale 1:48

Medium Tank T25E1

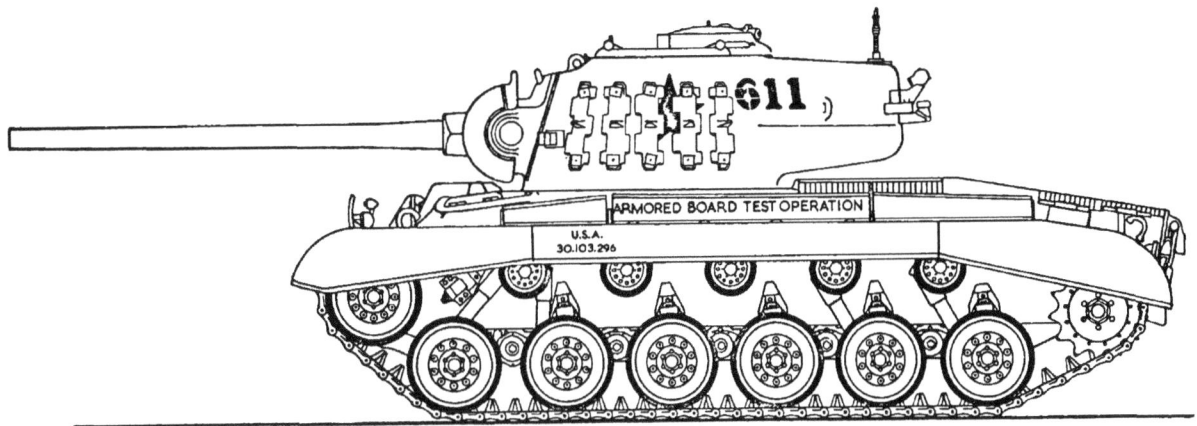

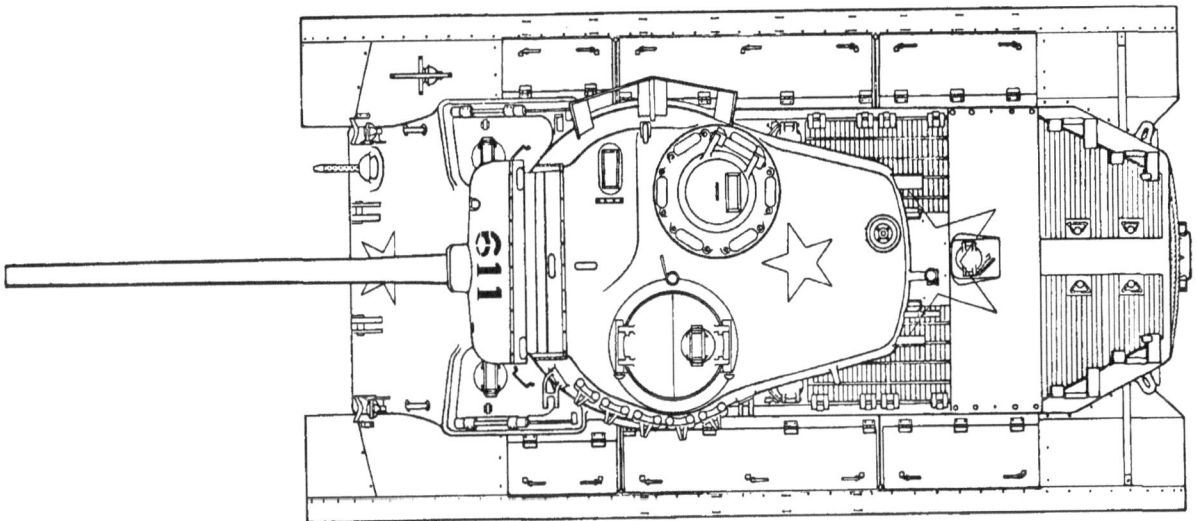

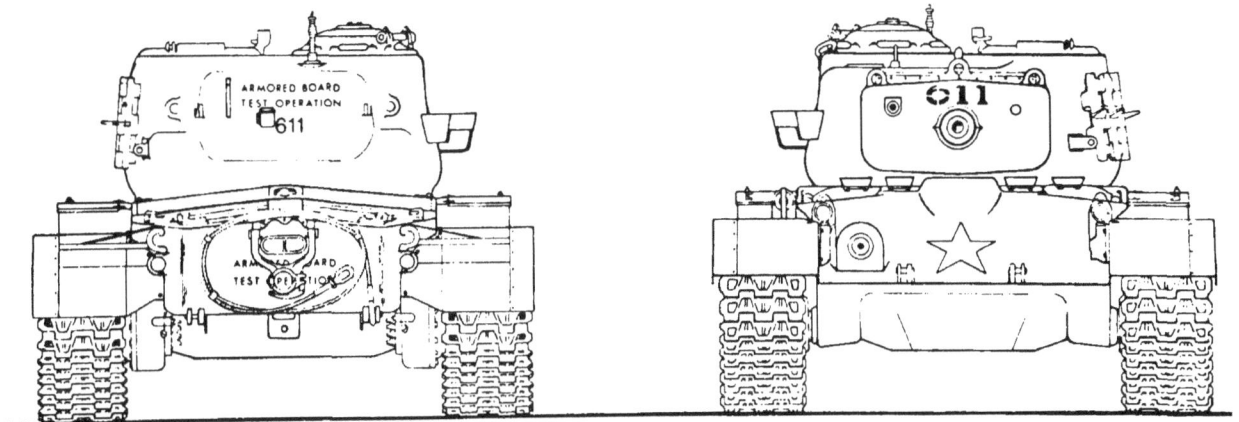

ARMORED BOARD TEST OPERATION

U.S.A.
30.103.296

611

Scale 1:48

Medium Tank T26E1

RIGHT SIDE OF VEHICLE

RIGHT FUEL TANK

GENERATOR AND AUXILIARY ENGINE

"LEFT" ANGLE DRIVE PROPELLER SHAFT

"LEFT" CYLINDER HEAD

RIGHT FANS

FINAL DRIVE

TRANSMISSION: PLANETARY REDUCTION GEARS TORQUE CONVERTER PLANETARY TRANSMISSION GEARS

RIGHT AIR CLEANER

"LEFT" MAGNETO (Viewed from drive end) Magneto Rotor (R)— Counterclockwise Distributor (D) Clockwise

EXHAUST INTAKE CAMSHAFTS

1L 2L 3L 4L

DIFFERENTIAL

FLYWHEEL

FRONT OF VEHICLE (MAGNETO END OF ENGINE)

D R

R D

REAR OF VEHICLE (FLYWHEEL END OF ENGINE)

CRANKCASE BREATHER

"RIGHT" MAGNETO (Viewed from drive end) Magneto Rotor (R)— Clockwise Distributor (D)— Counterclockwise

INTAKE EXHAUST CAMSHAFTS

4R 3R 2R 1R

REAR CARBURETOR

FRONT CARBURETOR

CRANKING MOTOR

NOT SHOWN: ROTATION OF— WATER PUMP— CLOCKWISE OIL FILTER MANUAL TURNING NUT— CLOCKWISE

LEFT AIR CLEANER

LEFT FUEL TANK

"RIGHT" CYLINDER HEAD

"RIGHT" ANGLE DRIVE PROPELLER SHAFT

FINAL DRIVE

LEFT SIDE OF VEHICLE

THROTTLE CONTROL ROD BELLCRANK — CARBURETOR GUARDS — CARBURETORS — RIGHT WATER MANIFOLD — THROTTLE CONTROL ROD

ROCKER ARM LINK — SPARK PLUG COVER

BOOSTER PISTON ROD ROCKER ARM — RIGHT EXHAUST MANIFOLD

GOVERNOR — RIGHT CYLINDER BLOCK DRAIN PLUG

ROCKER ARM STOP SCREW — RIGHT FAN DRIVE PROPELLER SHAFT

GOVERNOR BOOSTER — STARTER TERMINAL COVER

ENGINE THROTTLE BELLCRANK ROD — STARTER

ENGINE THROTTLE BELLCRANK — GROUND STRAP TERMINAL

THROTTLE CONTROL ROD LEVER LINK — RIGHT ENGINE LEG

SHOCK MOUNTING

CARBURETOR DEGASSERS — LEFT WATER MANIFOLD

RIGHT WATER MANIFOLD — ENGINE OIL FILLER

CARBURETOR ADAPTER — LEFT REAR HEATER TUBE

RIGHT EXHAUST MANIFOLD — LEFT CYLINDER HEAD

LEFT EXHAUST MANIFOLD — FAN DRIVE PROPELLER SHAFT UNIVERSAL JOINTS

FLYWHEEL PILOT BEARING — CYLINDER BLOCK

STARTER — ENGINE OIL PAN

FLYWHEEL

TRANSMISSION OIL LEVEL GAGE — TRANSMISSION BREATHER

CONTROLLED DIFFERENTIAL — DIFFERENTIAL BREATHER

UNIVERSAL JOINTS — FINAL DRIVE

DRIVE SPROCKETS

HUB AND FLANGE ASSEMBLY

PLANETARY GEAR SECTION

TORQUE CONVERTER — TRANSMISSION

REDUCTION GEAR UNIT

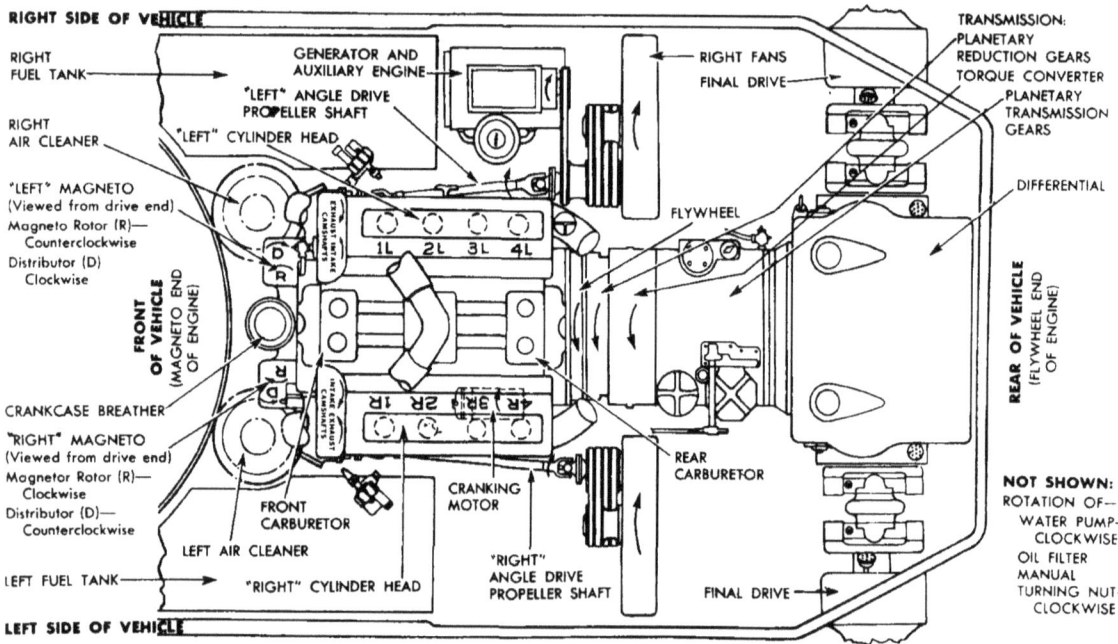

The arrangement of the engine compartment is diagramed above showing the location of the fuel tanks and power train components. Front (top left) and rear (center left) views of the Ford GAF engine are compared here. The flywheel housing bolted directly to the transmission reduction gear unit. The complete power train is shown in the lower photograph.

The power unit, consisting of the Ford GAF engine, torqmatic transmission, and controlled differential, was installed in the engine compartment and was removeable as a single assembly. Two fuel tanks were mounted, one on each side of the engine, at the forward end of the compartment. The cooling unit, removeable as a single assembly, was attached to the cast transverse housing, with the fans fitting on each side of the transmission.

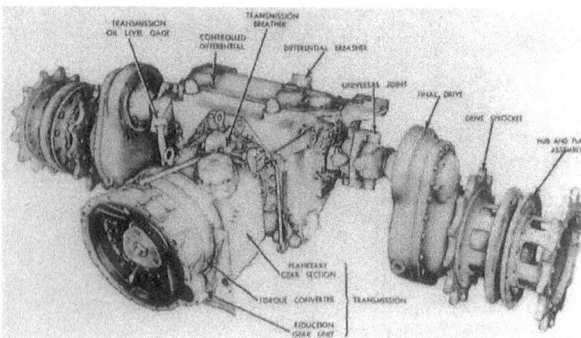

The model GAF engine, quite similar to the GAN used in the T23, was a 60 degree V-type, eight cylinder, four cycle, gasoline power plant developing 500 gross horsepower at 2600 rpm. The torqmatic transmission was an improved version of the model fitted to the T20 and T20E3. It consisted of the planetary reduction gears bolted to the engine flywheel, the torque converter which served as a fluid clutch, and the transmission planetary reduction gears which provided the three forward and one reverse gear ratios. Bolted to the rear cover of the transmission was the controlled differential which transmitted power through universal joints to the final drives and provided for steering and braking the vehicle.

112

The final drives were bolted to the outside of the hull side plates at the rear of the tank and transmitted power from the differential to the track sprockets. Each final drive consisted of a final drive pinion shaft, a pinion and gear, a final drive shaft, and a final drive housing. The final drive shaft was flanged at the outer end and provided with studs for mounting the track sprocket hubs. The gears and bearings inside the final drive housing operated in a continuous oil bath.

The power unit consisting of the engine, transmission, and controlled differential could be removed as a single assembly (above). At left, the remaining components are exposed in the engine compartment. The control rods and cables can be seen along the compartment floor. The driving controls are diagramed immediately below and at the bottom of the page, the driver's station is shown at the left and that of the assistant driver is at the right.

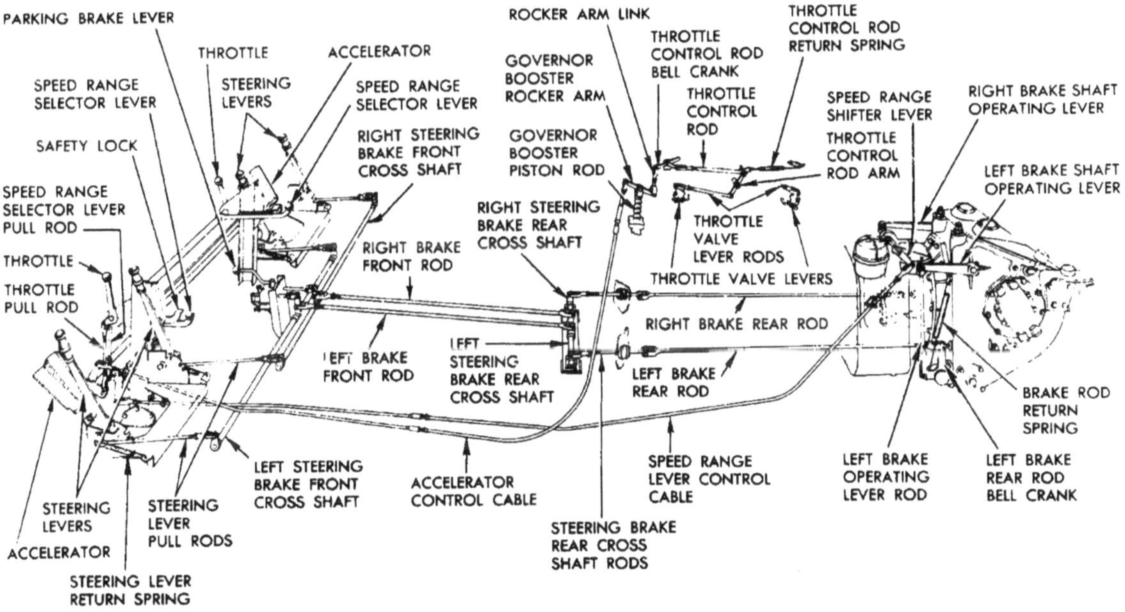

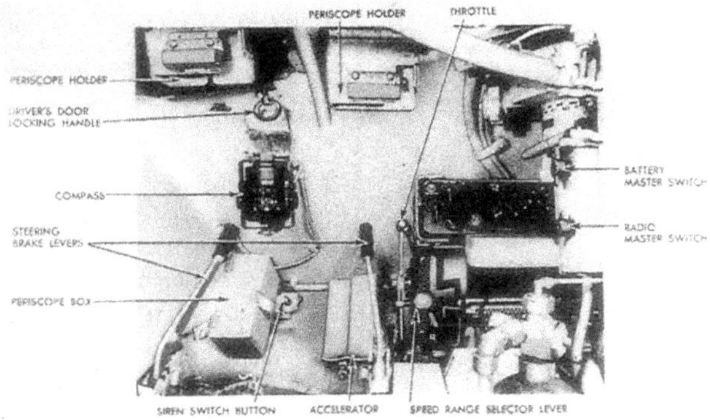

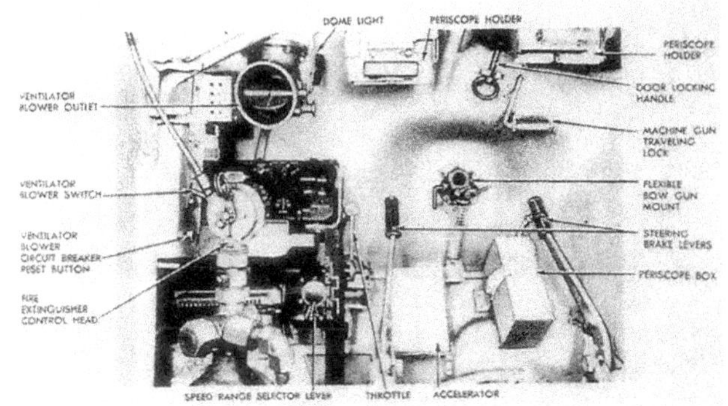

113

The boat shaped bottom of the Pershing hull is clearly revealed in this view from the lower rear of the tank. The engine compartment drain covers can be seen, as well as the drivers' escape hatches at the front end. The road wheel arm supports are bolted to the floor side plates.

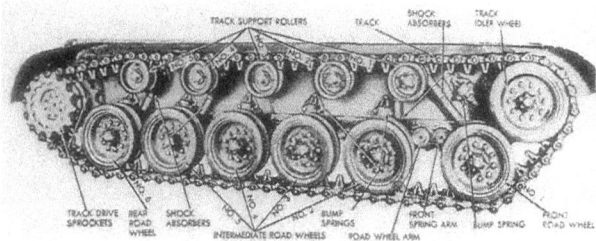

The details shown on the T25E1 suspension above apply equally to the T26E1 except for the wider tracks and heavier duty components of the latter vehicle.

In appearance the torsion bar suspension system was quite similar on the T25E1 and T26E1. However, the greater weight of the latter vehicle required heavier duty components and the 19 inch track width of the T25E1 was increased to 24 inches. Both tanks rode on 12, 26 inch diameter, rubber tired, dual road wheels, six for each track. Although the same diameter, the wheels for the T26E1 were wider and of heavier construction. Each dual road wheel was bolted to the road wheel hub mounted on two tapered roller bearings on each road wheel arm spindle. In turn, the road wheel arm was supported on two needle bearings in a road wheel arm support. These road wheel arm supports also provided seats for the torsion bar anchor plugs for the end of the torsion bar extending from the opposite side of the vehicle. The torsion bars, made of spring steel, ran crosswise from the road wheel arm to the anchor plug in the opposite road wheel arm support. Springing action to support the weight of the tank was obtained from the torsional resistance of the bar when twisted by the up and down movement of the road wheel arm and spindle. Volute type bumper springs acted as stops for each road wheel arm. Hydraulic shock absorbers controlled the movement of the first two and last two road wheels on each side of the tank. The track idler wheel was supported on an eccentric spindle in the upper end of the front road wheel arm. The spindle position could be moved to adjust the track tension. When the front road wheel struck an obstacle and moved upward, the arm pivoted on the front road wheel arm spindle moving the idler forward and down, thus maintaining the track tension. The upper section of each track was carried on five dual track support rollers mounted on brackets bolted to the hull side plates.

All steel, single pin tracks were used for both vehicles differing slightly in design. On the 19 inch wide T25E1 tracks, the sprocket engaged the lugs at each side of the track shoes. With the wider T26E1 tracks, the sprocket teeth entered holes at each end of the track shoe. There were 82 shoes per track with a six inch pitch for both tanks.

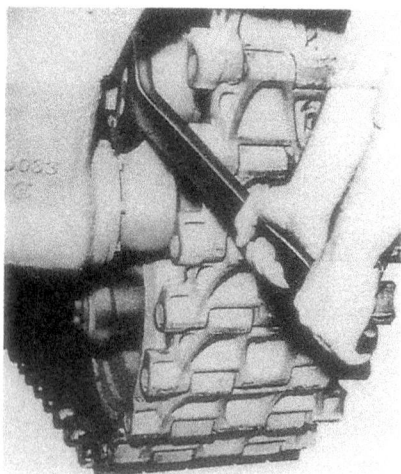

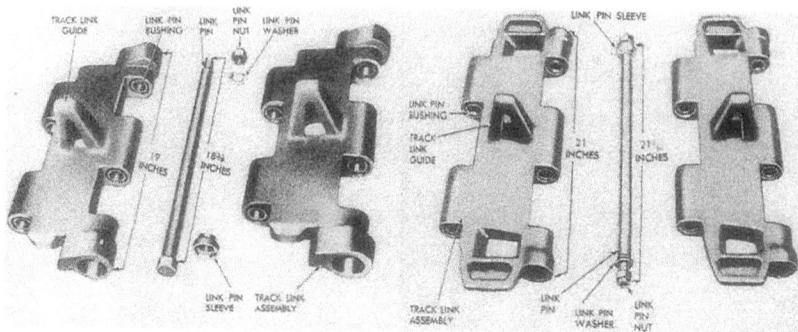

The track tension is being adjusted at left above. At right above, the single pin track link assemblies are compared for the T25E1 (left) and T26E1 (right). The overall width of the T26E1 link was 24 inches and this track was designated the T81.

114

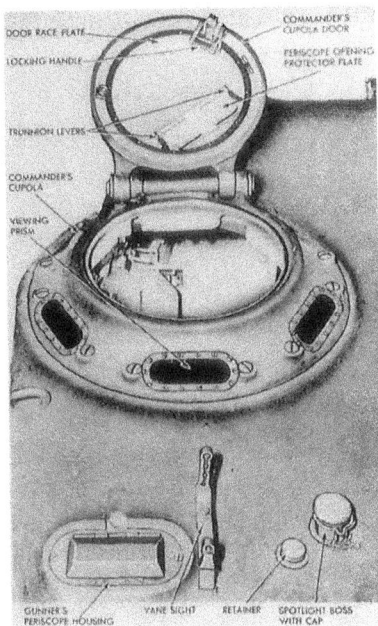

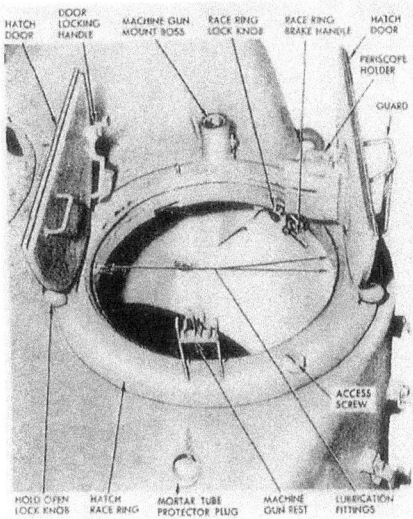

Views of the turret hatches from the front show the tank commander's cupola (top left) and the loader's rotating ring hatch (top center). Note the port for the 2 inch smoke mortar just in front of the loader's hatch. The turret bustle stowage just behind the loader's position (top right) illustrates the ready access to ammunition for the machine guns and smoke mortar. The opposite side of the turret bustle (below) contained the radio located next to the tank commander.

A distinguishing feature of both tanks was the heavy cast turret set well forward on the hull. It reflected many of the same design characteristics as the T23 which paralleled it in production, although heavier and larger to accommodate the 90 mm gun. The turret sides and rear were 2½ inches thick for the T25E1 compared to 3 inches for the T26E1. A semicircular rotating platform and a partial basket were suspended from the right side of the turret. The inside diameter of the turret ring was the same 69 inches as on the other T20 series tanks, as well as the Sherman. The large turret bustle needed to balance the long 90 mm gun provided stowage space for the radio equipment.

A six block vision cupola was mounted over the tank commander's position on the right side of the turret. In addition, the cupola hatch carried a standard M6 periscope in a rotating mount. In front of the tank commander, the gunner's M8A1 periscope with a built in telescope was fitted through the turret roof and coupled to the gun mount. The gunner was also equipped with an M71C telescope in the gun shield, coaxial with the 90 mm gun. At the left of the gun, a circular double door hatch was provided for the loader with a revolving ring mount for a .50 caliber antiaircraft machine gun. A rotating M6 periscope was located in one of the hatch doors. The port for the 2 inch smoke mortar M3, appeared just forward of the loader's hatch. The mortar, used for laying smoke screens, was fixed in the turret roof and traversed by rotation of the turret. Range could be set at approximately 35, 75, or 150 yards by use of the regulator on the mortar controlling the escape of propellant gases.

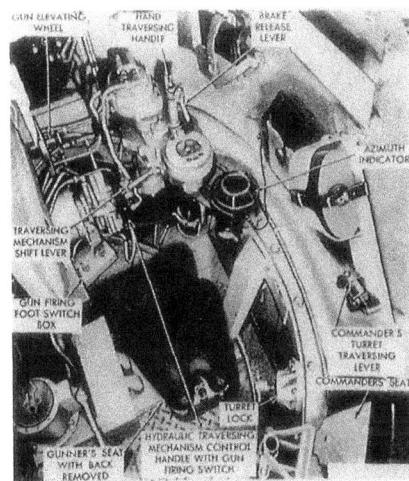

The gunner's station as viewed from the tank commander's hatch. Note the floor of the partial turret basket under the gunner's seat.

115

The T99 and T99E1 combination gun mounts were fitted to the T25E1 and T26E1 respectively. The only difference was a maximum armor thickness of 3½ inches for the T99 and 4½ inches for the T99E1. Both carried the 90 mm gun T7, with a .30 caliber machine gun M1919A4 in a coaxial mount just to the left of the main weapon. Elevation ranged from +20 to -10 degrees, but no stabilizer was fitted. Turret rotation provided a 360 degree traverse either manually or by a hydraulic drive. The 90 mm gun could be fired electrically by the trigger on the power traverse control handle or by a foot switch on the turret floor. A manual firing pedal was also provided in case of electrical failure. Another electric foot switch fired the coaxial machine gun, which could also be fired manually by the loader, using the conventional trigger.

Forty-two rounds of 90 mm ammunition were stowed in two water protected bins under the floor of the fighting compartment. Access to the ammunition required that the turret be traversed until the rotating semicircular platform exposed the correct bin. The cover could then be opened and the ammunition removed.

The gun mount recoil cylinders are visible in the view at right from the loader's position. Note the coaxial .30 caliber machine gun and the roof mounted 2 inch smoke mortar at the left of the photograph.

Above is the 90 mm gun T7 and below the same weapon is shown in the combination gun mount T99E1.

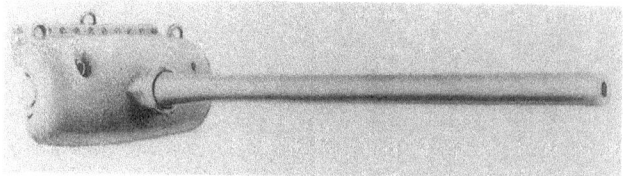

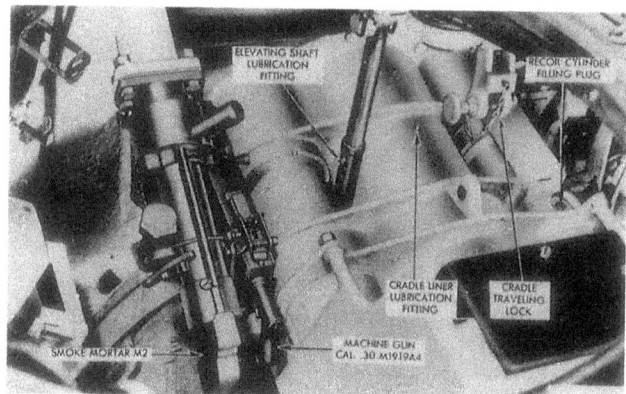

The floor stowage compartments for the 90 mm ammunition are fully exposed below with the turret removed. The 90 mm ammunition was placed in compartments B, C, D, and E. Details of the water tanks which provided the "wet stowage" are shown in the two views at the left. These tanks acted as dividers in the ammunition compartments and were also placed at each end along the side walls of the tank hull.

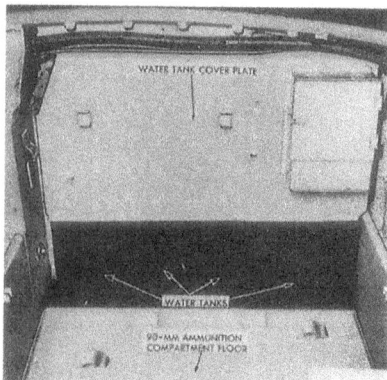

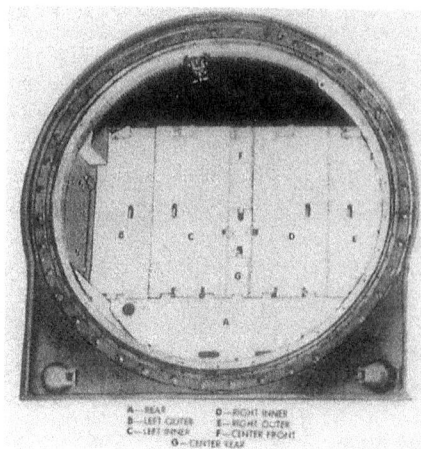

With the arrival of the first tanks at Aberdeen and Fort Knox, an intensive test program began to discover any defects in the vehicles and make any necessary changes. One result of these tests was a change in the steering ratio from 2.08:1 to 1.79 to 1. The change increased the minimum turning diameter by about 20 feet.

The initial testing of the first and second T26E1s was completed on May 21, 1944 and Aberdeen concluded that the vehicle design was basically sound and that the torqmatic transmission, torsion bar suspension, and 90 mm gun were satisfactory. Major problems were poor cooling of the differential, repeated failures of the connection between the engine and flywheel, plugging of the radiators cores with dirt, oil leaks, and poor 90 mm ammunition stowage. The latter also raised serious objections at Fort Knox. First, the 42 rounds carried were considered insufficient as the ETO wanted at least 70 rounds and second, the stowage bins under the turret platform were far too inaccessible. Tests also revealed that the semicircular rotating platform under the gunner and tank commander interfered with the loader during firing. Several solutions to the problem were considered, one of which eliminated the assistant driver and replaced him with an 18 round ammunition rack. However, the ETO ruled this out by insisting on the retention of the bow machine gun. The final design removed the rotating platform and turret basket, retaining only a small foot rest for the gunner. The floor ammunition racks were enlarged and the water protection sacrificed to obtain additional ammunition space. Racks were installed on each side of the fighting compartment as well as a ready rack on the loader's side of the turret. With the new arrangement, 70 rounds of 90 mm ammunition could be carried.

Firing tests at Fort Knox showed a severe obscuration problem with the 90 mm gun. So much dust was kicked up by the muzzle blast that the 90 was condemned as a one shot weapon, because of the long delay before it was possible to see well enough to fire again. The cure recommended was a muzzle brake which would deflect the blast sideways reducing the obscuration along the line of sight.

Ground pressure on the T25E1 was considered too high with the narrow 19 inch tracks. Armored Board tests showed the necessity for wider tracks and it was recommended that the complete suspension of the T26E1 be installed. In addition to the 24 inch wide tracks, the heavier duty components were expected to greatly reduce maintenance requirements. However, by the summer of 1944, interest had shifted to the T26E1 with its heavier armor plate and no production was approved for the T25E1. Allied troops in Normandy were running into increasing numbers of heavily armored enemy fighting vehicles and were calling for more firepower and greatly increased protection. This combat situation killed any further interest in the T25E1 and stimulated the development of even more heavily armored vehicles even though the horsepower to weight ratio suffered accordingly. The desire for a heavy tank even led to some name changing. On June 29, 1944, all models of the T26 series were redesignated as heavy tanks and remained so until after the war, when the original designation of a medium tank was restored. Throughout this narrative they will be considered as medium tanks.

In January 1944, Ordnance had received permission to build 250 T26E1s beyond the initial ten pilot tanks. The lessons learned from the pilot test program were applied to the production of these vehicles beginning in November at the Fisher Tank Arsenal. With these improvements, the tank was redesignated as the T26E3. The serial numbers of the 250 tanks were 11 through 260, since serials 1 through 10 were applied to the T26E1. The registration numbers were 30119821 through 30120070.

To check the various modifications, T25E1 number 5 (registration number 30103256) was converted to the T26E3 standard. The turret was modified to the latest design practice by replacing the revolving hatch over the loader's position with a small oval hatch. The .50 caliber machine gun was moved

T25E1 number 5 modified to T26E3 standard. The 90 mm gun is fitted with a muzzle brake and is carried in the combination mount T99E2.

to a pedestal mount at the rear of the turret roof. The combination mount T99E2 was installed carrying the 90 mm gun with a double baffle muzzle brake. This mount was later standardized as the M67. The turret basket was removed and the stowage arrangements reworked to the new standard. Brackets were welded to the turret which allowed the installation of a hoisting device for removing the power train.

The modified T25E1 number 5 with the gun locked in the travel position. A steel plate has been welded into the opening for the large rotating loader's hatch and a small oval hatch installed.

Eclipsed by the desire for heavier armor, the T25E1 was limited to use as an experimental vehicle and some served in this role even into the postwar period. The 90 mm gun T14 in the combination gun mount T102 was tested in the T25E1 during May and June 1945. This mount used a concentric recoil mechanism which occupied much less space than the standard double recoil cylinders. As the name implied the recoil chamber encircled the gun tube itself replacing the separate cylinders spaced about the circumference of the tube. With the increased space available, two coaxial .50 caliber machine guns replaced the single .30 caliber weapon on the standard mount. The 90 mm gun T14 was fitted with a muzzle brake and was ballistically identical with the standard M3 gun.

T25E1 number 1 (at left) with the 90 mm gun T14 in concentric recoil mount T102. The .50 caliber machine gun barrels protruded through the gun shield (bottom left) and frequently served as hand holds for the driver when entering or leaving the tank. The arrangement of the two coaxial guns and the telescopic sight in the T102 mount are shown above. To provide space for the .50 caliber machine gun ammunition, the 90 mm ready rack (below) was limited to 8 rounds.

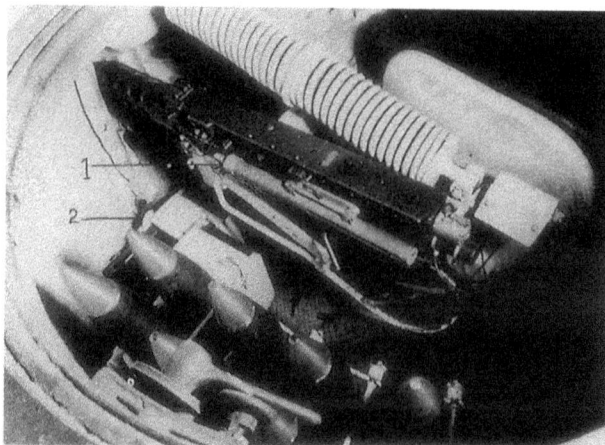

T25E1 number 13 with the integrated fighting compartment. The modifications required a new turret roof of increased height to provide space for the rangefinder. Eyepieces and controls for the rangefinder T31 are shown in the interior view below.

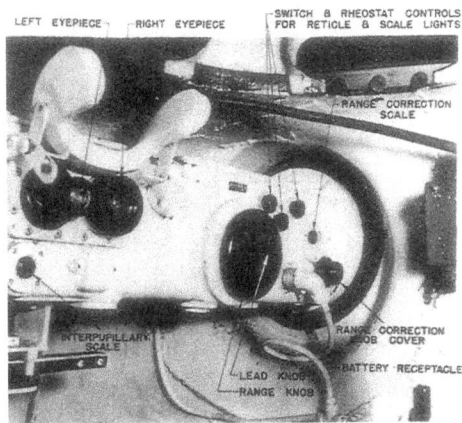

As late as 1948, T25E1 number 13 (registration number 30103264) was used to study the advantages of an integrated fighting compartment equipped with a turret range finder and a lead computer. These studies led directly to the development of the more sophisticated postwar fire control systems. Many other experimental combinations of armament were proposed and one proposal drawing even shows the 105 mm gun T5E1 mounted on a T25E1 tank with an enlarged turret ring.

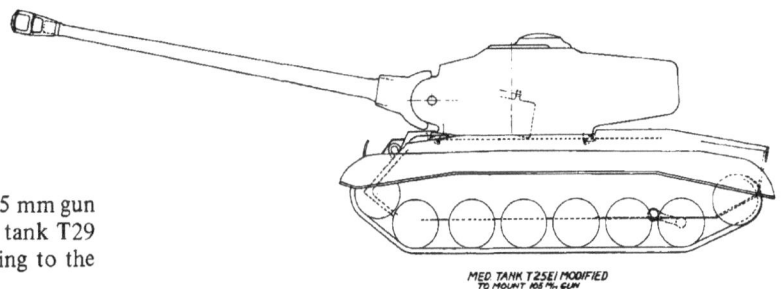

A proposal drawing for the T25E1 fitted with the 105 mm gun T5E1. This is the same weapon installed on heavy tank T29 and the turret ring has been enlarged in this drawing to the same 80 inches used on the T29.

MED TANK T25E1 MODIFIED
TO MOUNT 105 M. GUN

119

The second production T26E3 (serial number 12) at Aberdeen Proving Ground (above and below) on December 12, 1944.

After correcting most of the problems uncovered by the test program, the T26E3 went into production at the Fisher Tank Arsenal in November 1944 and the first 10 vehicles were delivered by the end of the month. Production rapidly increased with 30 vehicles delivered in December, 70 in January and 132 in February 1945. In March, production also started at the Detroit Tank Arsenal and the combined delivery for the month was 194 tanks. This increased to 269 in April, 361 in May and 370 in June. By this time, the end of the war in Europe had reduced the production schedule, although over 2000 were produced by the end of 1945. After its successful introduction into battle, the T26E3 was standardized as the M26 in March 1945. The first vehicles were equipped with the 24 inch T81 track of the T26E1, but later production was fitted with the 23 inch T80E1 track, similar to that on the M4 with the horizontal volute spring suspension.

The 90 mm gun T7, mounted in the T25 and T26 series vehicles, was standardized as the 90 mm gun M3. This weapon was developed from the 90 mm antiaircraft gun modified for use in a tank mount. Its ballistic performance was identical to that of the antiaircraft gun and it used the same ammunition. A muzzle velocity of 2650 ft/sec was obtained with explosive loaded, armor piercing, capped ammunition (APC M82). This was later increased to 2800 ft/sec by using a larger powder charge, but few of these rounds reached the troops in time for use during World War II. Improved performance was obtained using armor piercing composite rigid (APCR) ammunition consisting of a lightweight shot with a tungsten carbide core. These projectiles, designated HVAP-T 90 mm T30E16, achieved a muzzle velocity of 3350 ft/sec and would penetrate 6.1 inches of homogeneous armor at 2000 yards and 30 degrees obliquity. Despite this performance, the highly sloped

T26E3 serial number 204 is shown in the three upper photographs at the right during tests at the General Motors Proving Ground. This tank was produced at the Fisher Tank Arsenal in January 1945 and is still equipped with the T81 track. Note the stowed position of the .50 caliber machine gun pedestal.

frontal armor of the German Panzerkampfwagen V Panther presented a difficult target. The M3 gun using T30E16 shot could penetrate the Panther's glacis plate only up to a dangerously close range of 450 yards. To help solve this problem, a shot was developed particularly for such high obliquity targets. This was the 90 mm AP-T shot T33 which was manufactured from the substitute standard monobloc shot M77. The M77 projectile was reheat treated to a higher hardness level and fitted with a windshield to improve its exterior ballistics. With the T33 shot, the M3 gun had a muzzle velocity of 2800 ft/sec, but it could penetrate the Panther's glacis up to a range of 1100 yards. Limited quantities of both the T30E16 and the T33 ammunition were available for the Pershings introduced by the Zebra mission. The effectiveness of both types in action has been described earlier.

Parallel with the T26E3's introduction to battle by the Zebra mission, 20 tanks were shipped to Fort Knox for test. These tests revealed that most problems with the T26E1 had been solved. However, one serious defect remained, a dangerously weak elevation housing on the 90 mm gun. A stiffener plate was added as a quick fix in production until a stronger forged housing could be procured.

Tests both at Fort Knox and Aberdeen showed that the ventilation provided by the rotoclone blower was inadequate to clear powder fumes from the tank when firing. A 1000 cfm blower was recommended to replace the 400 cfm unit. The change required a new front casting to house the larger fan. Previous ballistic tests had shown the bulge over the rotoclone somewhat weaker than the rest of the front plate. In the design of the new front casting, this area was thickened to improve protection in this region. The new casting was introduced into production with tank number 550 at Fisher and number 235 at Chrysler. At the same time, a study was started to consider remounting the ventilator in the rear of the turret bustle. However, this did not come about until the postwar medium tank T42. Research continued into improved cooling, better ballistic protection for air intakes and grilles, and better vision devices. Particular attention was given to the problem of increasing the ballistic protection of the gun shield, which was considerably weaker than the rest of the front armor. However, only minor changes were made in the standard M26 shield before production ended, such as a reduction in the clearance between the shield and the gun barrel to minimize jamming by small arms fire.

The M26 below is serial number 1602 produced at Fisher. This tank is equipped with the 1000 cfm rotoclone blower. The difference in the front armor casting is apparent when compared with serial number 204 above.

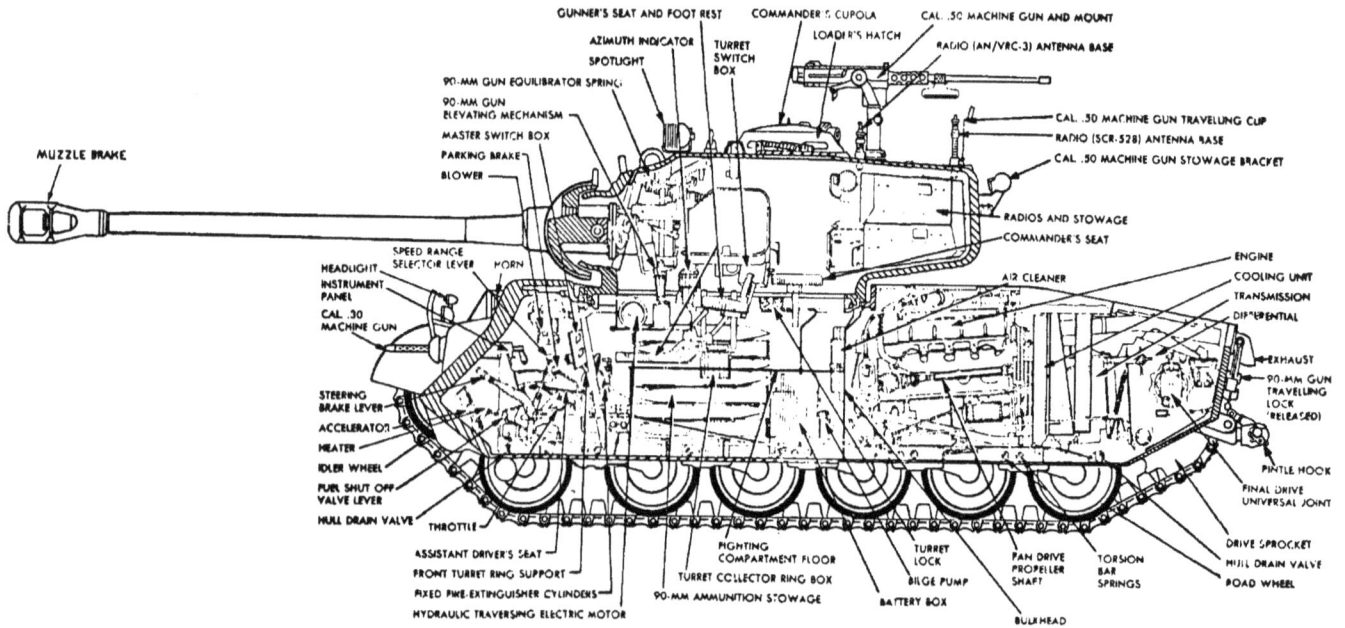

Sectional view of the production M26. Compare the increased ammunition stowage with the earlier version on page 109.

The production M26 in the two photographs below is fitted with the T80E1 double pin track. Details of this track can be seen on the four blocks attached to the tank turret.

Attempts to improve the performance in mud led to variations in the design of the track blocks. A modified T81 single pin track is under test below. The T91 double pin track with larger grousers is shown above.

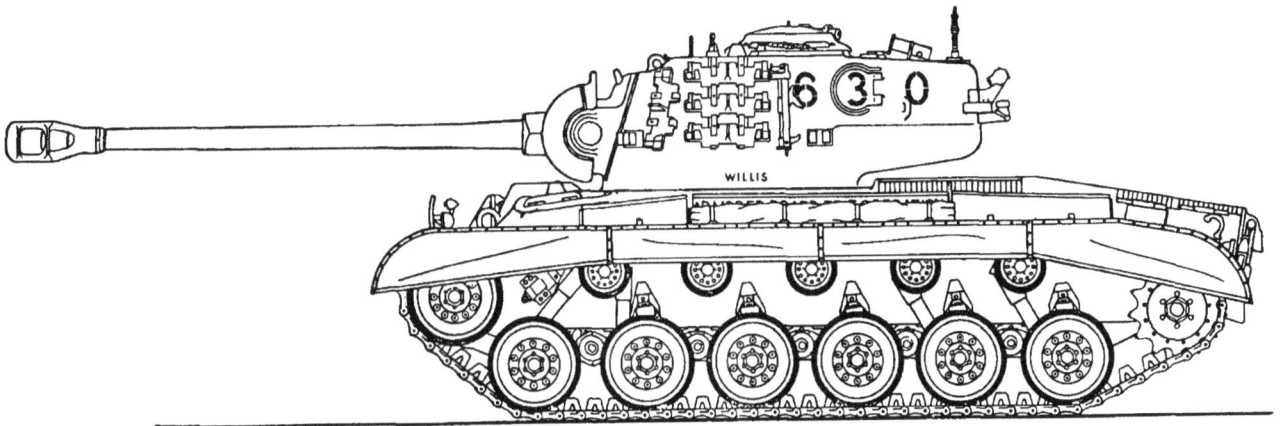

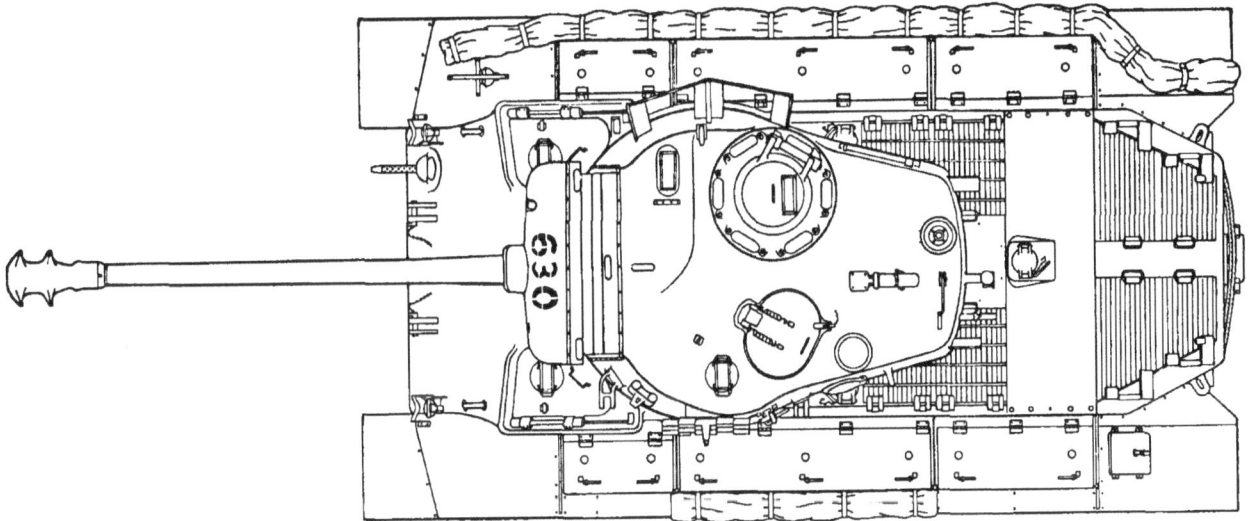

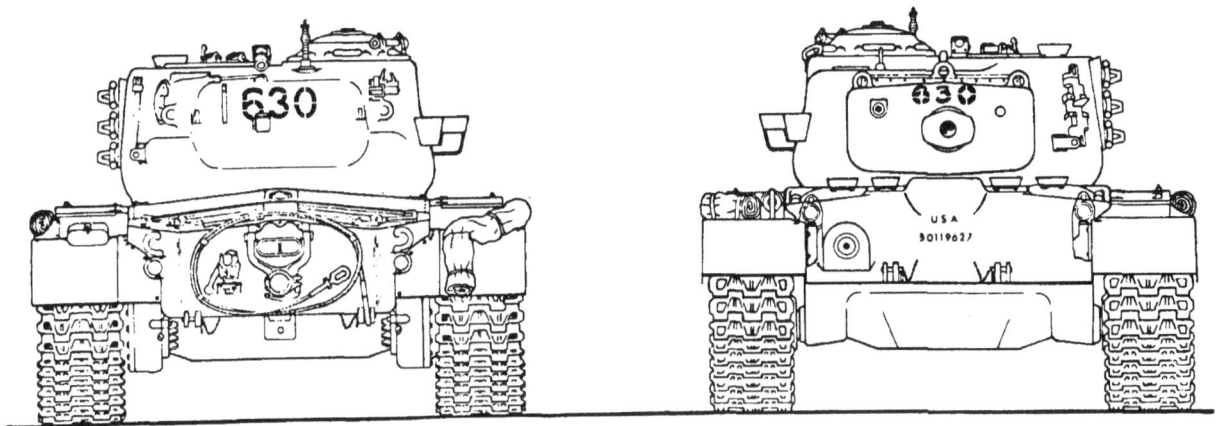

WILLIS

630

630

630

630

USA
30119627

Scale 1:48

Medium Tank M26 (T26E3)

123

The outer finished surface of road wheel arm supports must be parallel to C/L of tank

Surface within track support roller area must be within 3/32 inch

Differential mounting bracket

Normal top of hull

36 1/8
42 3/8 + 1/8
26 7/8 left side
30 11/16 right side
C/L turret
4 1/2 ± 1/32
20 11/16 ± 1/16
C/L of differential
18 7/16
C/L
For final drive opening

Under side of 1/2 floor
62 3/4 L
67 9/16 R
Engine mounting bracket

93 L
96 13/16 R

122 1/4 L
126 1/16 R

151 1/2 L
155 5/16 R

The finished face of these holes must be perpendicular to C/L of differential within 0.015 full indicator reading in 14-3/4 diam.

Center line turret
41 13/16
107 3/4
Center line outriggers #
10
62 11/16
Center line outriggers #3
Center line outriggers #4
Center line front suspension
Center line outriggers #2
Center line differential

35 7/16 ± 1/16
Bulkhead
83
40 ± 1/32
65 3/4 ± 1/16

9 1/2 ± 1/32
15.437 ± 1/16
2 1/4
11 7/8
55 1/8
35 1/4
32 7/8 ± 1/32

8 3/4 ± 1/64
12 7/8 ± 1/32

72 1/4
125
12 1/2 ± 1/32

Hull roof plate must be checked with serviceable turret race ring for flatness

Rear support
Front support
Reinforcement mounting support

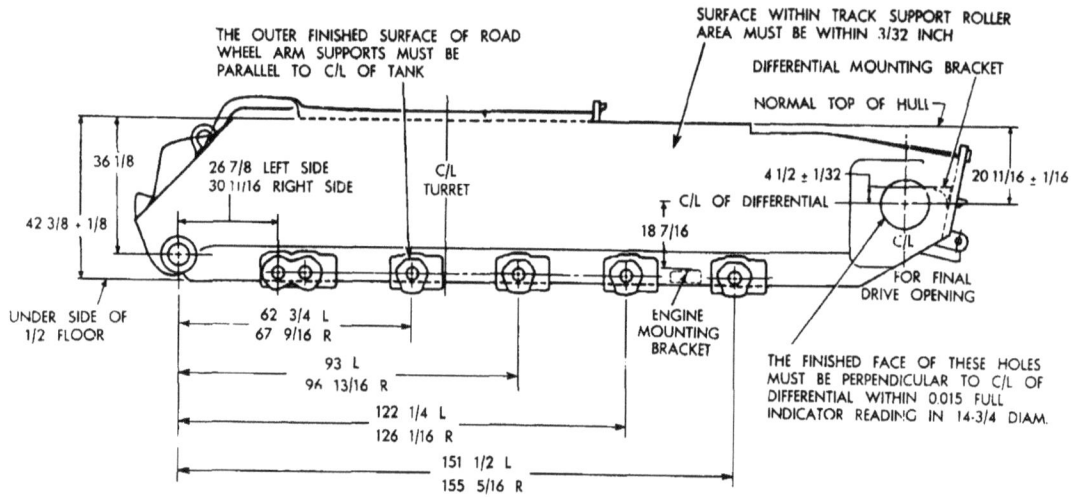

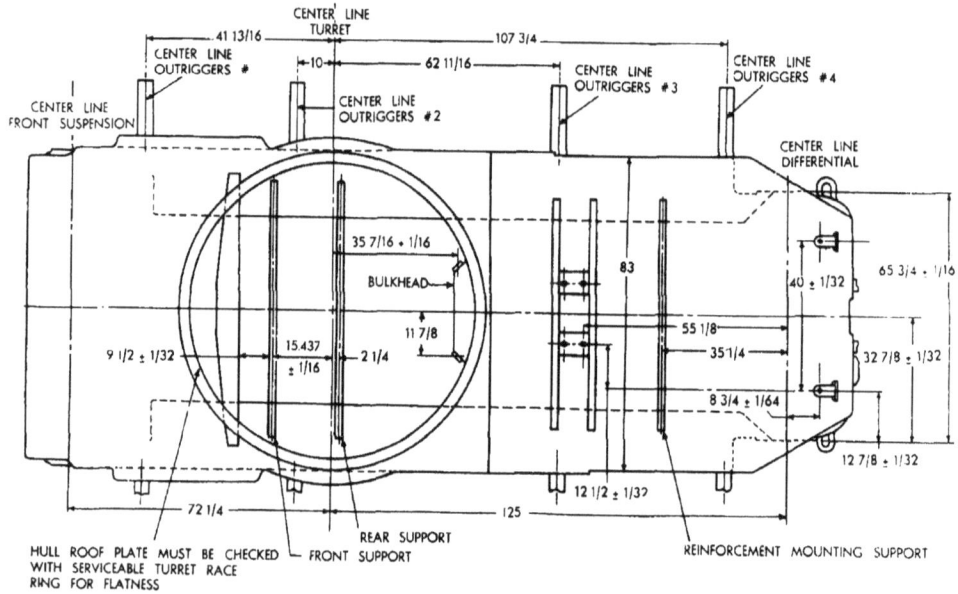

The dimensions of the production M26 hull are shown in these drawings. The welding diagram (below left) indicates the manner in which the hull was assembled from the armor castings and rolled sections.

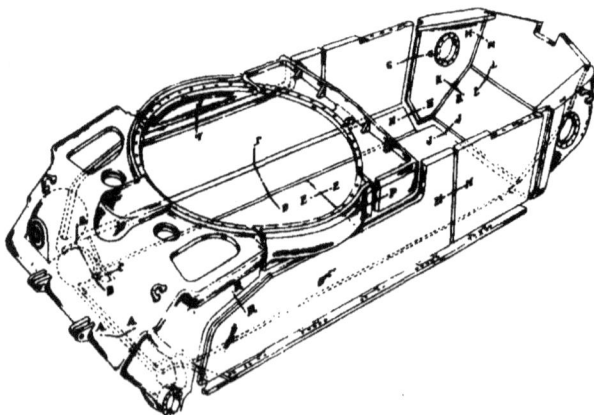

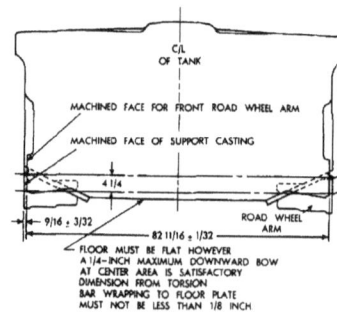

C/L of tank
Machined face for front road wheel arm
Machined face of support casting
4 1/4
9/16 ± 3/32
82 11/16 ± 1/32
Road wheel arm

Floor must be flat however a 1/4-inch maximum downward bow at center area is satisfactory dimension from torsion bar wrapping to floor plate must not be less than 1/8 inch

124

Production line of Pershings (right) at the Fisher Tank Arsenal during the winter of 1944-45.

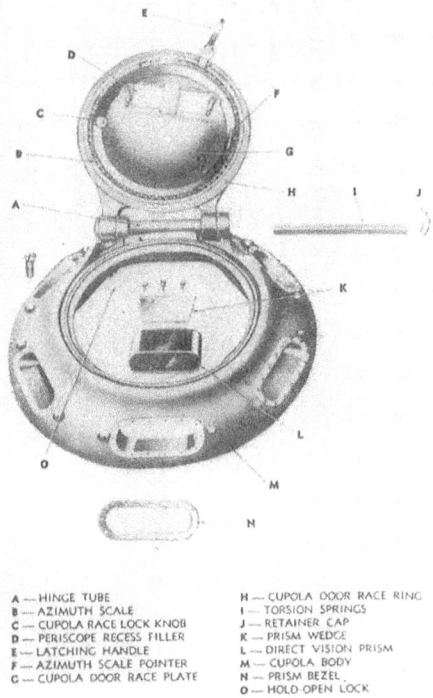

A — HINGE TUBE
B — AZIMUTH SCALE
C — CUPOLA RACE LOCK KNOB
D — PERISCOPE RECESS FILLER
E — LATCHING HANDLE
F — AZIMUTH SCALE POINTER
G — CUPOLA DOOR RACE PLATE
H — CUPOLA DOOR RACE RING
I — TORSION SPRINGS
J — RETAINER CAP
K — PRISM WEDGE
L — DIRECT VISION PRISM
M — CUPOLA BODY
N — PRISM BEZEL
O — HOLD-OPEN LOCK

Details of the commander's vision cupola are shown above. The other views show the relative location of items on the top of the tank. The bottom photographs detail the loader's hatch in the open and closed positions.

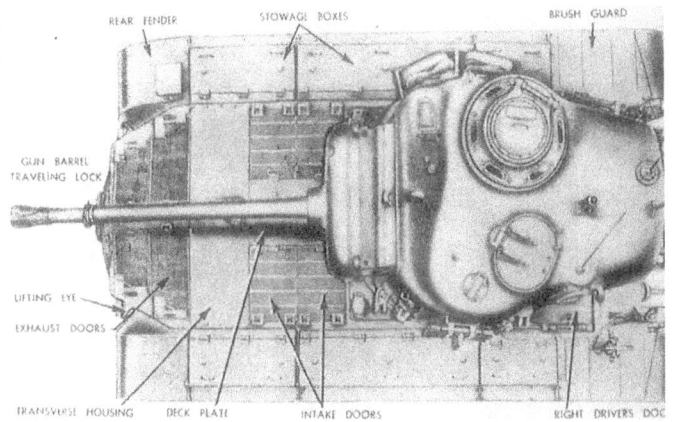

REAR FENDER
STOWAGE BOXES
BRUSH GUARD
GUN BARREL TRAVELING LOCK
LIFTING EYE
EXHAUST DOORS
TRANSVERSE HOUSING
DECK PLATE
INTAKE DOORS
RIGHT DRIVERS DOO

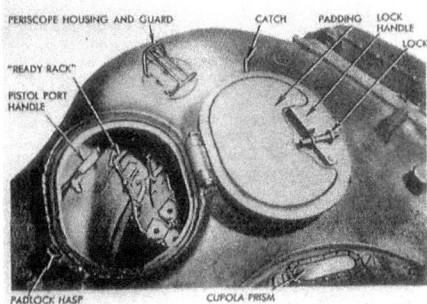

PERISCOPE HOUSING AND GUARD
CATCH
PADDING
LOCK HANDLE
LOCK
"READY RACK"
PISTOL PORT HANDLE
PADLOCK HASP
CUPOLA PRISM

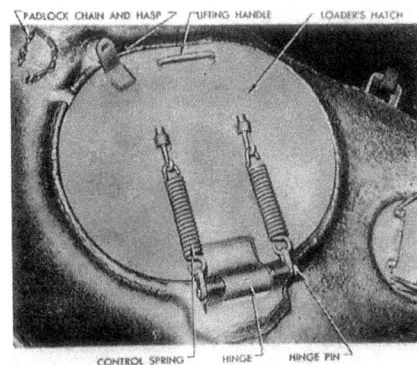

PADLOCK CHAIN AND HASP
LIFTING HANDLE
LOADER'S HATCH
CONTROL SPRING
HINGE
HINGE PIN

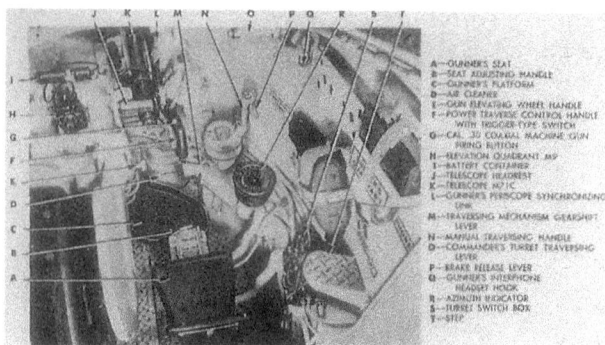

A—GUNNER'S SEAT
B—SEAT ADJUSTING HANDLE
C—GUNNER'S PLATFORM
D—AIR CLEANER
E—GUN ELEVATING WHEEL HANDLE
F—POWER TRAVERSE CONTROL HANDLE
  WITH TRIGGER-TYPE SWITCH
G—CAL .30 COAXIAL MACHINE GUN
  FIRING BUTTON
H—ELEVATION QUADRANT M9
I—BATTERY CONTAINER
J—TELESCOPE HEADREST
K—TELESCOPE M71C
L—GUNNER'S PERISCOPE SYNCHRONIZING
  LINK
M—TRAVERSING MECHANISM GEARSHIFT
  LEVER
N—MANUAL TRAVERSING HANDLE
O—COMMANDER'S TURRET TRAVERSING
  LEVER
P—BRAKE RELEASE LEVER
Q—GUNNER'S INTERPHONE
  HEADSET HOOK
R—AZIMUTH INDICATOR
S—TURRET SWITCH BOX
T—STEP

The two top photographs give a general view of the gunner's station. Note the replacement of the turret basket by the small foot rest marked C. Compare with the T26E1 on page 115. The general arrangement of the turret is shown in the drawing below.

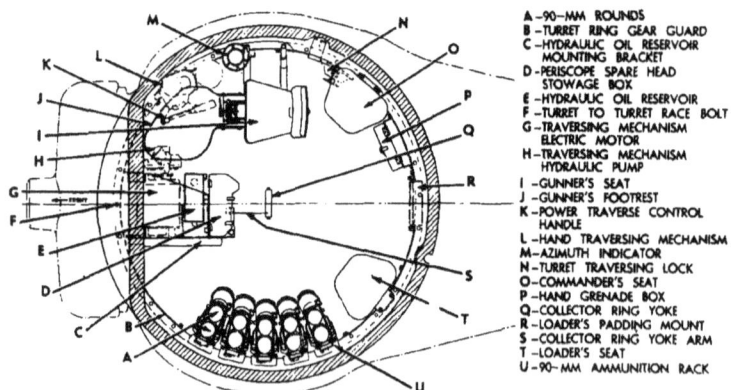

A—90-MM ROUNDS
B—TURRET RING GEAR GUARD
C—HYDRAULIC OIL RESERVOIR
  MOUNTING BRACKET
D—PERISCOPE SPARE HEAD
  STOWAGE BOX
E—HYDRAULIC OIL RESERVOIR
F—TURRET TO TURRET RACE BOLT
G—TRAVERSING MECHANISM
  ELECTRIC MOTOR
H—TRAVERSING MECHANISM
  HYDRAULIC PUMP
I—GUNNER'S SEAT
J—GUNNER'S FOOTREST
K—POWER TRAVERSE CONTROL
  HANDLE
L—HAND TRAVERSING MECHANISM
M—AZIMUTH INDICATOR
N—TURRET TRAVERSING LOCK
O—COMMANDER'S SEAT
P—HAND GRENADE BOX
Q—COLLECTOR RING YOKE
R—LOADER'S PADDING MOUNT
S—COLLECTOR RING YOKE ARM
T—LOADER'S SEAT
U—90-MM AMMUNITION RACK

A—COMPASS
B—PERISCOPE
C—DIRECT SIGHT TELESCOPE
D—CAL .30 MACHINE GUN ELEC-
  TRICAL FIRING TRIGGER
E—MANUAL TURRET TRAVERSE
  CONTROL HANDLE
F—90-MM GUN ELECTRICAL
  FIRING TRIGGER
G—HYDRAULIC TURRET TRAVERSE
  CONTROL HANDLE
H—ELEVATING HANDWHEEL
J—TURRET TRAVERSE GEARSHIFT
  LEVER

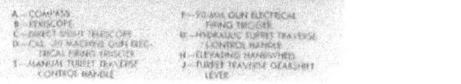

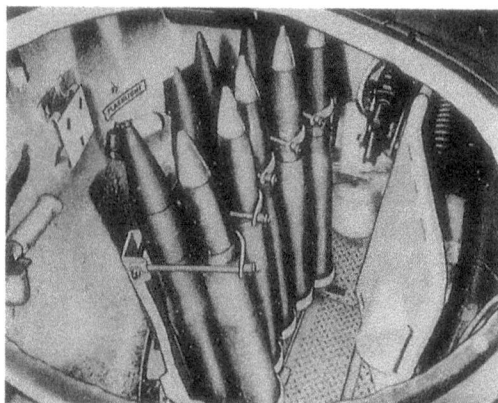

The components of the SCR528 radio were located (above) in the turret bustle just behind the tank commander.

The ready rack (at left) provided ten rounds of 90 mm ammunition for immediate use at the loader's position.

With the turret removed, the floor stowage compartments can be seen open and closed in the bottom photographs.

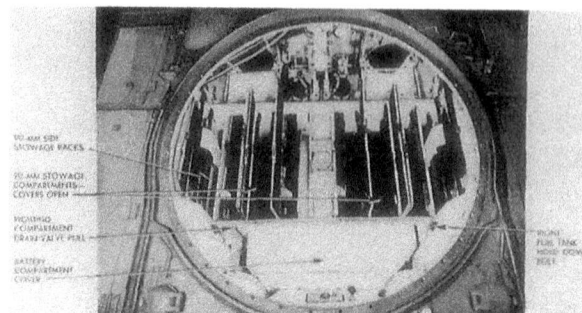

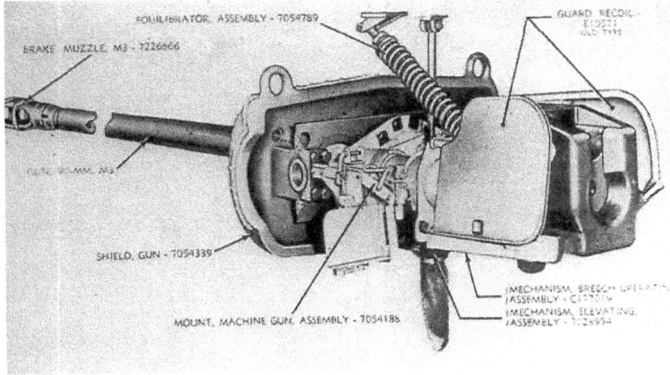

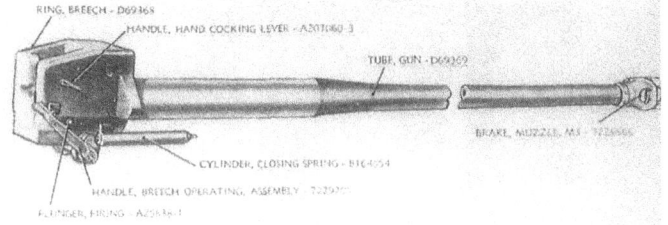

The 90 mm gun M3 is shown at the above left in combination mount M67. The .30 caliber coaxial machine has not been fitted in this photograph. The 90 mm gun removed from the mount appears above at the right.

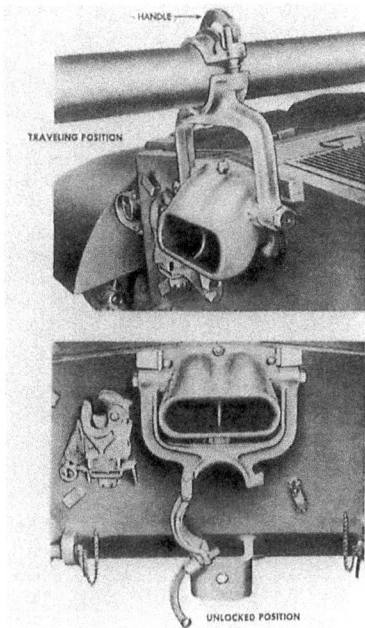

Above is a closeup of the muzzle brake M3. The insert in the second baffle is replaceable when eroded.

The traveling lock for the 90 mm gun is shown in both positions at the left. Also, note the towing pintle in the stowed position.

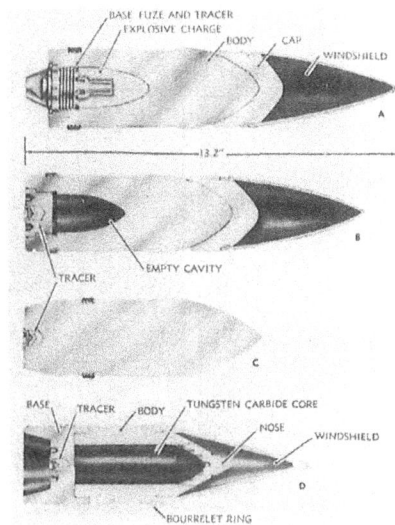

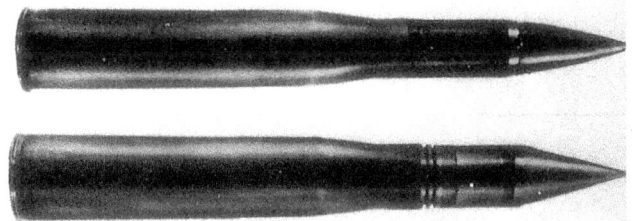

HVAP T30E16 (APCR-T)   AP T33 (APBC-T)

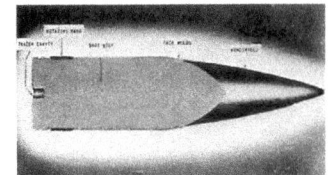

Projectile types are shown at the left from top to bottom, (A) APC (APCBC/HE-T), (B) APC (APCBC-T), (C) AP (AP-T), and (D) HVAP (APCR-T). The two complete 90 mm rounds in the upper photograph at right are AP T33 (APBC-T) at the top and HVAP T30E16 (APCR-T) at the bottom. Details of the projectiles for these rounds are shown in the lower pictures.

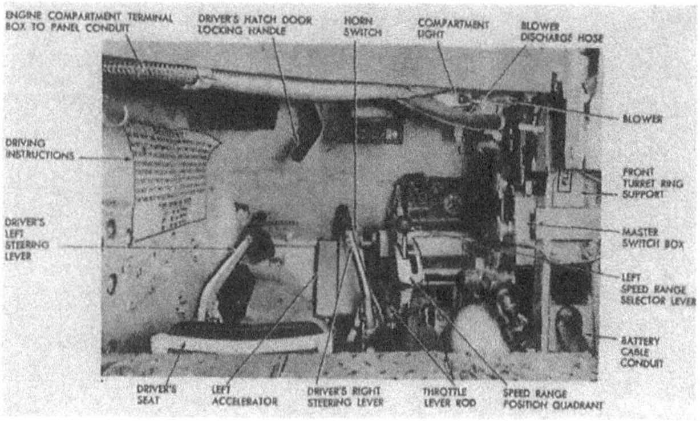
ENGINE COMPARTMENT TERMINAL BOX TO PANEL CONDUIT
DRIVER'S HATCH DOOR LOCKING HANDLE
HORN SWITCH
COMPARTMENT LIGHT
BLOWER DISCHARGE HOSE
BLOWER
DRIVING INSTRUCTIONS
FRONT TURRET RING SUPPORT
DRIVER'S LEFT STEERING LEVER
MASTER SWITCH BOX
LEFT SPEED RANGE SELECTOR LEVER
BATTERY CABLE CONDUIT
DRIVER'S SEAT
LEFT ACCELERATOR
DRIVER'S RIGHT STEERING LEVER
THROTTLE LEVER ROD
SPEED RANGE POSITION QUADRANT

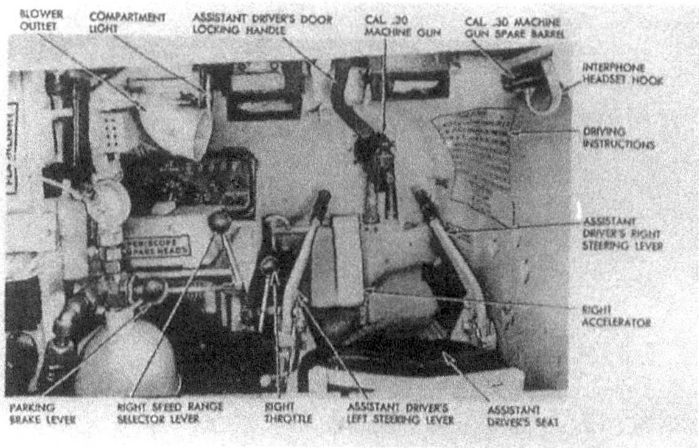
BLOWER OUTLET
COMPARTMENT LIGHT
ASSISTANT DRIVER'S DOOR LOCKING HANDLE
CAL .30 MACHINE GUN
CAL .30 MACHINE GUN SPARE BARREL
INTERPHONE HEADSET HOOK
DRIVING INSTRUCTIONS
ASSISTANT DRIVER'S RIGHT STEERING LEVER
RIGHT ACCELERATOR
PARKING BRAKE LEVER
RIGHT SPEED RANGE SELECTOR LEVER
RIGHT THROTTLE
ASSISTANT DRIVER'S LEFT STEERING LEVER
ASSISTANT DRIVER'S SEAT

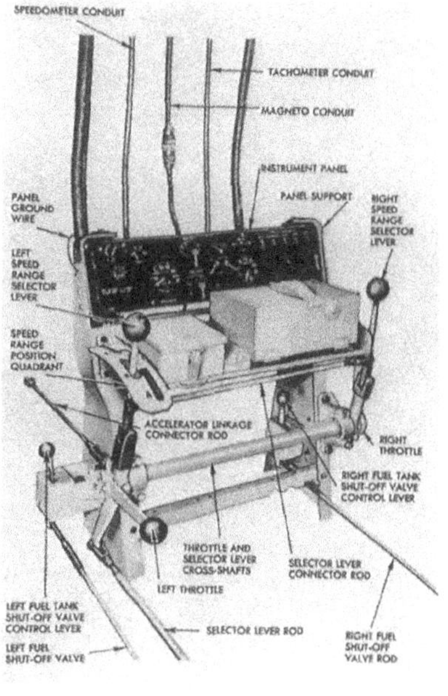
SPEEDOMETER CONDUIT
TACHOMETER CONDUIT
MAGNETO CONDUIT
INSTRUMENT PANEL
PANEL SUPPORT
PANEL GROUND WIRE
RIGHT SPEED RANGE SELECTOR LEVER
LEFT SPEED RANGE SELECTOR LEVER
SPEED RANGE POSITION QUADRANT
ACCELERATOR LINKAGE CONNECTOR ROD
RIGHT THROTTLE
RIGHT FUEL TANK SHUT-OFF VALVE CONTROL LEVER
THROTTLE AND SELECTOR LEVER CROSS-SHAFTS
SELECTOR LEVER CONNECTOR ROD
LEFT THROTTLE
LEFT FUEL TANK SHUT-OFF VALVE CONTROL LEVER
SELECTOR LEVER ROD
LEFT FUEL SHUT-OFF VALVE
RIGHT FUEL SHUT-OFF VALVE ROD

Above: The driver's position at the left and the assistant driver's position at the right.
The control console is shown at the left with a closeup view of the instrument panel below.

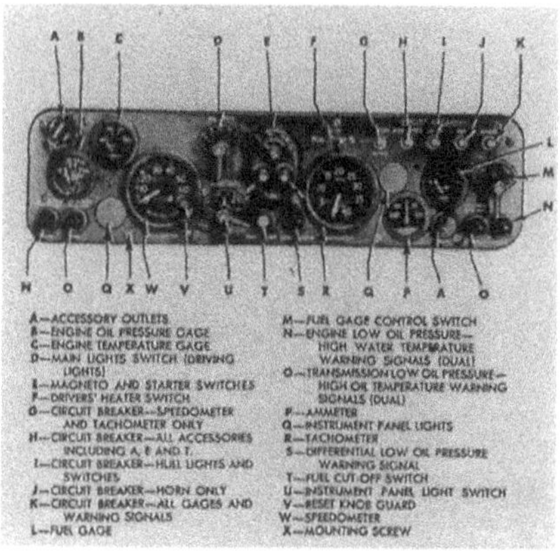
A—ACCESSORY OUTLETS
B—ENGINE OIL PRESSURE GAGE
C—ENGINE TEMPERATURE GAGE
D—MAIN LIGHTS SWITCH (DRIVING LIGHTS)
E—MAGNETO AND STARTER SWITCHES
F—DRIVERS' HEATER SWITCH
G—CIRCUIT BREAKER—SPEEDOMETER AND TACHOMETER ONLY
H—CIRCUIT BREAKER—ALL ACCESSORIES INCLUDING A, F AND T.
I—CIRCUIT BREAKER—HULL LIGHTS AND SWITCHES
J—CIRCUIT BREAKER—HORN ONLY
K—CIRCUIT BREAKER—ALL GAGES AND WARNING SIGNALS
L—FUEL GAGE
M—FUEL GAGE CONTROL SWITCH
N—ENGINE LOW OIL PRESSURE—HIGH WATER TEMPERATURE WARNING SIGNALS (DUAL)
O—TRANSMISSION LOW OIL PRESSURE—HIGH OIL TEMPERATURE WARNING SIGNALS (DUAL)
P—AMMETER
Q—INSTRUMENT PANEL LIGHTS
R—TACHOMETER
S—DIFFERENTIAL LOW OIL PRESSURE WARNING SIGNAL
T—FUEL CUT-OFF SWITCH
U—INSTRUMENT PANEL LIGHT SWITCH
V—RESET KNOB GUARD
W—SPEEDOMETER
X—MOUNTING SCREW

OUTSIDE VIEW
CAL .30 MACHINE GUN, M1919A4

Above: The bow machine gun operated by the assistant driver.
Left: These two views show details of the drivers' hatches.
Below: One of the two escape hatches located in front of the drivers' seats.

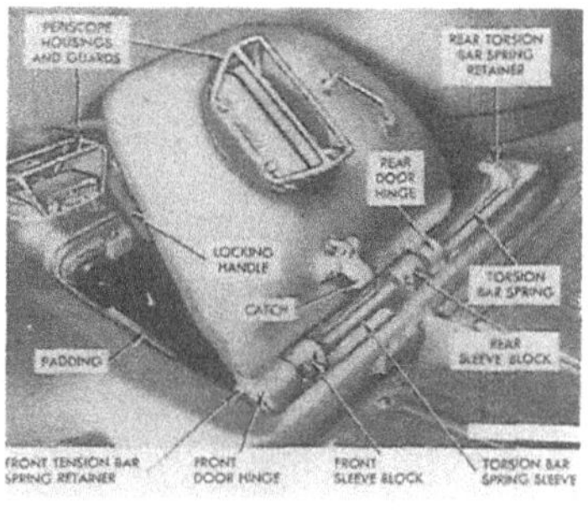
PERISCOPE HOUSINGS AND GUARDS
REAR TORSION BAR SPRING RETAINER
REAR DOOR HINGE
LOCKING HANDLE
TORSION BAR SPRING
CATCH
REAR SLEEVE BLOCK
PADDING
FRONT TENSION BAR SPRING RETAINER
FRONT DOOR HINGE
FRONT SLEEVE BLOCK
TORSION BAR SPRING SLEEVE

LOCK AND HANDLE
PERISCOPE HOUSING
CATCH RELEASE
EXTERIOR FIRE-EXTINGUISHER PULL HOUSING AND SPLASH SHIELD
HORN AND GUARD
PADDING

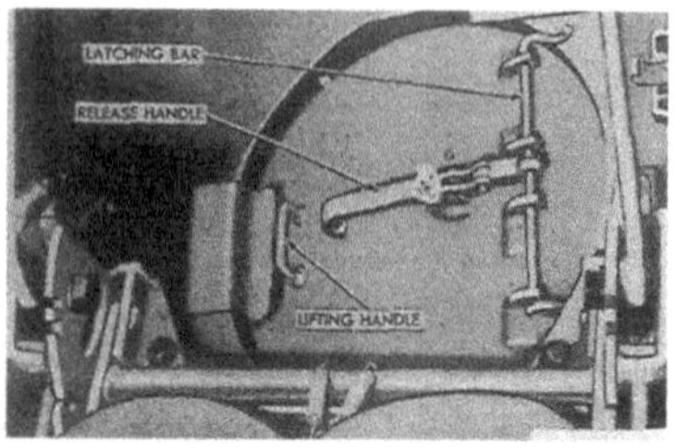
LATCHING BAR
RELEASE HANDLE
LIFTING HANDLE

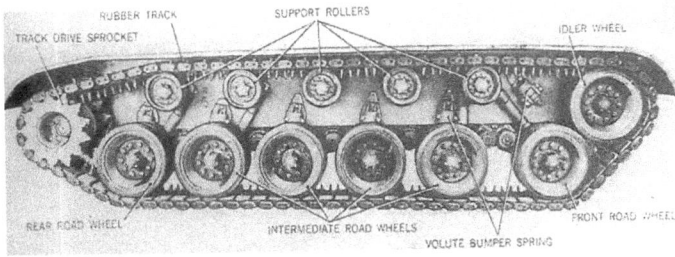

Above: M26 suspension fitted with the T80E1 double pin tracks. The links for the T80E1 track are shown below. Compare with the single pin track on page 114.

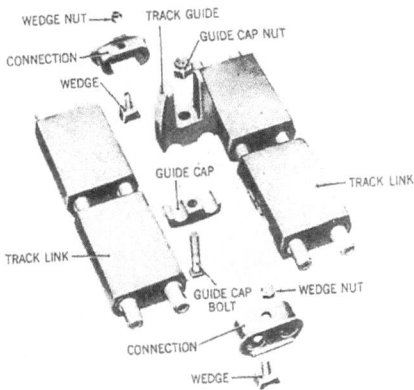

Above: Production line assembly of the M26 at Fisher Tank Arsenal reveals details of the suspension system. The shock absorbers and volute bumper springs are clearly visible.

The turret mounted hoisting device for removing the power unit is demonstrated at the right on T26E1 number 1.
Below: Attempts to increase the range of the Pershing resulted in experiments with jettisionable fuel tanks. As shown in these photographs, the T25E1 was again used as the test vehicle.

Two 18 ton inflatable bridge pontons are used above to float the Pershing.

During its service career, the Pershing was modified and fitted with a great variety of equipment to improve its mobility and firepower. Considerable effort was expended on improving the deep water fording ability and for amphibious operations, swimming devices consisting of jettisonable flotation tanks were developed. However, the latter were extremely bulky and could hardly have been transported on shipboard where space was at a premium.

Two views of the M26 prepared for deep water fording are shown at the right and below.

Further development of the waterproofing technique for deep water fording is apparent in the two top photographs. Note the overall cover used to seal around the gun shield in comparison with the method illustrated on the opposite page. The telescope and coaxial machine gun ports also remain open allowing the weapons to be fired. The two large brackets fixed to each side of the hull next to the first and last road wheels are attachment points for flotation gear. Details of the engine compartment covers are shown in the two photographs at the right. The covers provided a watertight base for the large cooling air stacks. Note the modified version of the travel lock for the 90 mm gun.

Two stages in the sealing procedure for the gun shield are illustrated below. The use of sealant around the drivers' hatches and periscopes is also clearly visible.

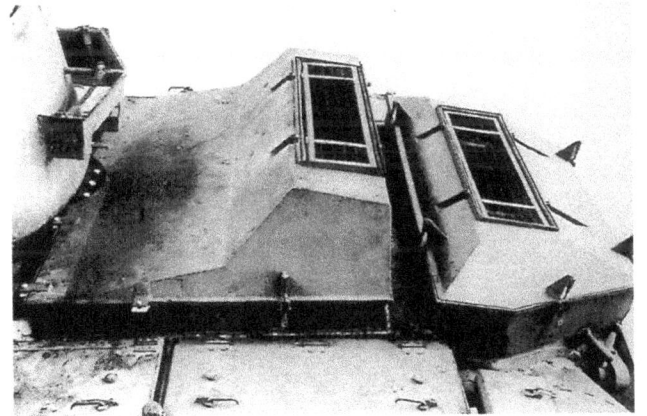

M26 serial number 2011 equipped with the T99 rocket launcher at Chrysler. This was the 250th Pershing produced at Detroit Tank Arsenal and it is fitted with the T80E1 tracks.

Efforts to increase the tanks firepower led to the installation of the 4.5 inch multiple rocket launcher T99 consisting of two launchers, one on each side of the turret with a combined total of 44 rockets. The launchers were traversed by rotating the turret and, by means of a connection to the gun shield, were elevated with the 90 mm gun. The launchers were mounted on trunnions and fitted with a device that allowed them to be jettisoned from inside the tank. The rockets could be fired individually, in salvos of two, or by full automatic ripple fire. With the latter, all 44 rockets could be fired in about seven seconds.

Above: Pershing, serial number 2011, under test at Aberdeen Proving Ground in December 1945.
Right: Closeup of the T99 rocket launcher showing the linkage to the gunshield.

Experience during the war indicated a need for more secondary armament to deal with soft targets. The standard machine gun mounts were considered insufficient and ways were sought to increase the secondary firepower. Some experiments mounted machine guns over the fenders which were fired by remote control from inside the tank. Another approach replaced the commander's cupola with a twin machine gun turret. This was designated the twin machine gun mount T121 and was designed to fit on either the M26 or M4 medium tanks. The mount was armed with either two .30 caliber or two .50 caliber machine guns and was aimed and fired from inside the turret. The mount was tested at Aberdeen on both the Pershing and the Sherman in April 1946 along with the fender mount guns.

Right: A comparison between the T121 mount and the standard vision cupola. The extra height of the T121 was a major disadvantage of this type of mount.
Below: The T121 mount with two .50 caliber machine guns under test on the Pershing at Aberdeen Proving Ground. Note the .30 caliber machine gun mounted on the right fender.

T26E2 production pilot (above and below) at Aberdeen Proving Ground in July 1945.

THE M45 (T26E2), T26E4, AND T26E5

Because of the successful application of the 105 mm howitzer to the M4 tank series, early consideration was given to arming the T23 with the same weapon. Following the "combat team" concept, it was desirable to use a common chassis for as many weapons as possible. A combination mount, based on the one used with the 76 in the T23, was designed and built for the 105 mm howitzer M4. With the shift in interest to the T26E1, the design was reworked, incorporating a heavier gun shield, for application to that vehicle. Mounting the howitzer, the tank was assigned the designation T26E2. Drawings of the howitzer mount, turret, and fighting compartment were released to the Fisher Tank Arsenal and Chrysler at Detroit Tank Arsenal in October 1944. Wooden mock-ups were prepared to aid in the design and the necessary stowage rearrangement to replace the 90 mm ammunition with 105 mm rounds. The gun shield was redesigned to minimize any impact loading on the trunnions and trunnion bearings which might result from a hit on the shield. The first turret was constructed at Fisher for installation on the pilot chassis built at Detroit Tank Arsenal.

The 105 mm howitzer M4 was considerably lighter in weight than the 90 mm gun it replaced and this was compensated by increasing the maximum thickness of the gun shield to 8 inches and the front turret armor to 5 inches. The turret sides varied from 3 to 5 inches in thickness. The new mount carried the coaxial .30 caliber machine gun M1919A4, an M76G telescope, and was designated the combination mount T117, later standardized as the M71. Hydraulic power traverse was provided and the mount was stabilized in

elevation which ranged from -10 to +35 degrees. The rearranged stowage allowed space for 74 rounds of 105 mm ammunition. Other features of the tank were identical with the late production M26.

Although scheduled for completion in April 1945, interest in the 105 mm howitzer as tank armament had begun to drop and the pilot tank (registration number 30131576) was not delivered to Aberdeen until July. Originally, it was intended to produce more tanks armed with the 105 mm howitzer, than with the 90 mm gun. However, later battle experience in Europe had reversed this trend by the spring of 1945 and major emphasis was centered on the high velocity gun. Both Chrysler and Fisher were originally issued contracts to produce the T26E2. With the end of the war in Europe and reduced interest in the howitzer as tank armament, Fisher's contract was cancelled and Chrysler's reduced in numbers. Production began in July 1945 at Detroit Tank Arsenal. Production was cut back further at the end of the war and only 185 vehicles were produced by the end of 1945. In the postwar period the tank was standardized as the medium tank M45 and some saw action in Korea.

The T26E2 production pilot shows several modifications standardized for late production Pershings. Note the additional pedestal for the .50 caliber machine gun in front of the commander's cupola and the exhaust muffler mounted on the rear of the hull.

Two views of the T26E2 during Armored Board tests at Fort Knox are shown above and at the left. Note the low position of the coaxial machine gun compared to the 90 mm gun Pershings. On this tank, the .50 caliber machine gun has been stowed, fully assembled, on the turret bustle rack. The barrel can be seen protruding from behind the turret.

The gun mount cover is removed in the photograph at the right revealing the heavy curved gun shield. This shield was almost twice as thick as that on the standard Pershing M67 gun mount, reaching a maximum of eight inches. Like other late production Pershings, the T26E2s were equipped with the T80E1 double pin track.

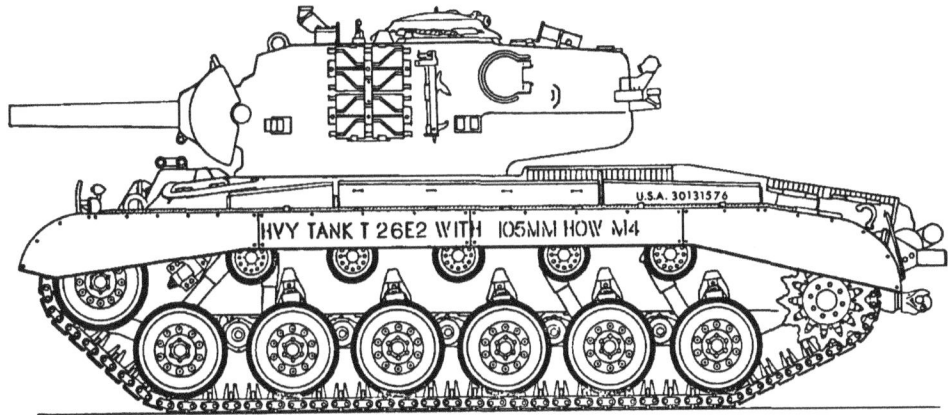

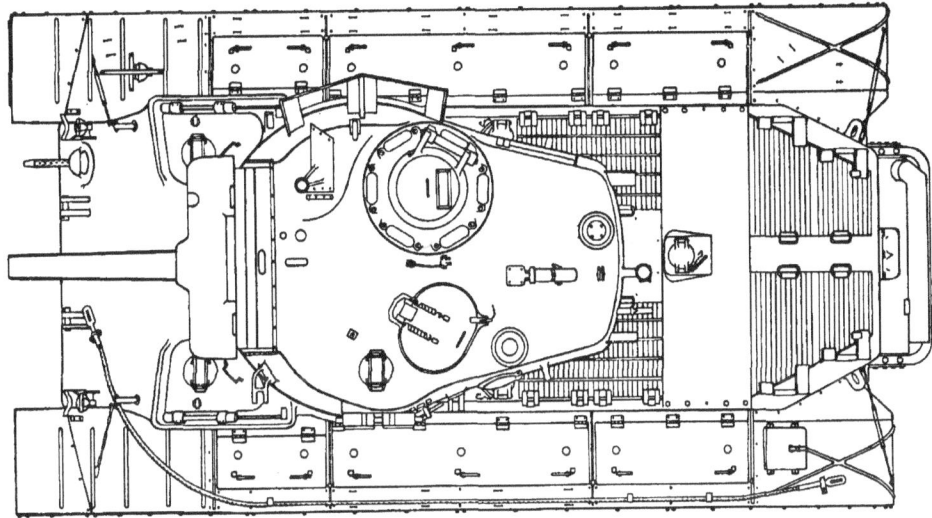

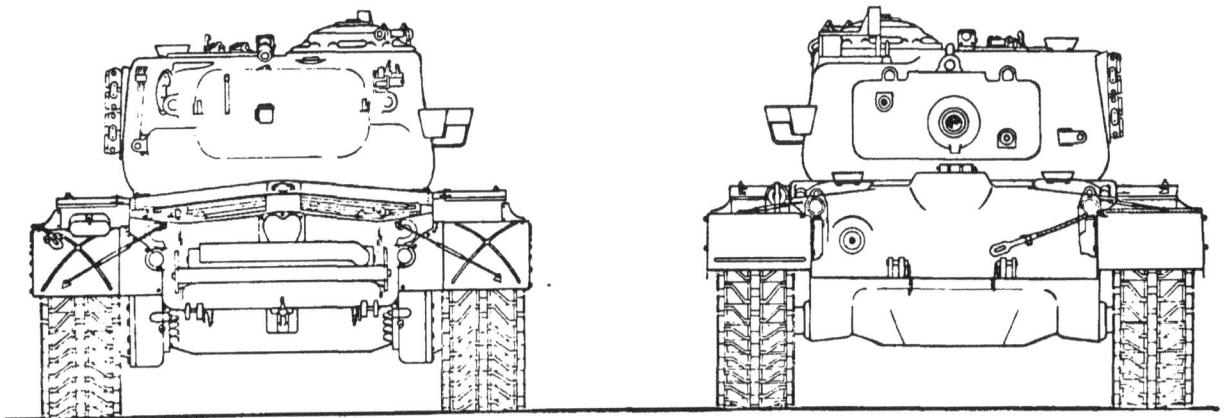

HVY TANK T 26E2 WITH 105MM HOW M4

U.S.A. 30131576

Scale 1:48

Medium Tank M45 (T26E2)

MOUNT FOR CAL. .50 HB MACHINE GUN M2
SEALED COVER FOR COMBINATION GUN MOUNT M25
105-MM HOWITZER M4
CAL. .30 MACHINE GUN M1919A4

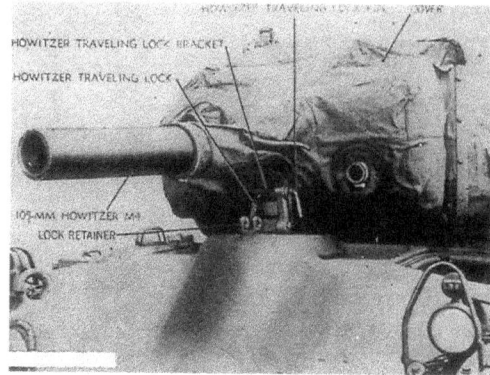

HOWITZER TRAVELING LOCK BRACKET
HOWITZER TRAVELING LOCK
HOWITZER TRAVELING LOCK COVER
105-MM HOWITZER M4
LOCK RETAINER

BREECH OPERATING HANDLE
ELEVATION QUADRANT M9
HAND COCKING LEVER HANDLE
HAND COCKING HANDLE STOP PIN
PERCUSSION MECHANISM
BREECHBLOCK
BREECH RING
CAL. .30 MACHINE GUN M1919A4

Armament details of the M45 include the howitzer traveling lock and coaxial .30 caliber machine gun port (above right). The hull mounted drivers' periscopes have been eliminated leaving only one periscope per driver, fitted in the hatch cover. The interior view from the loader's position (at left) shows the breech of the 105 mm howitzer M4 and the coaxial .30 caliber machine gun.

The inside of the turret shows the gunner's station (at left) and the radio equipment stowed in the turret bustle just back of the loader. The AN/VRC-3 radio was used for communication with infantry units during support operations.

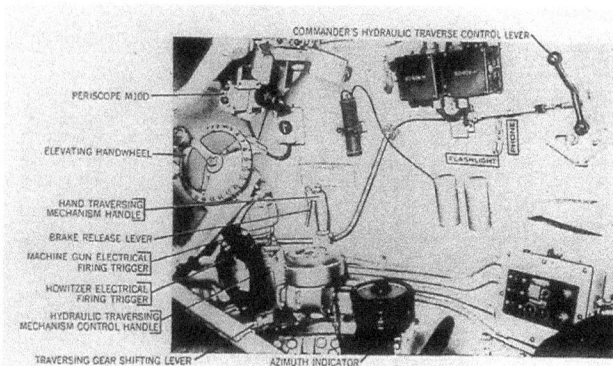

COMMANDER'S HYDRAULIC TRAVERSE CONTROL LEVER
PERISCOPE M16D
ELEVATING HANDWHEEL
HAND TRAVERSING MECHANISM HANDLE
BRAKE RELEASE LEVER
MACHINE GUN ELECTRICAL FIRING TRIGGER
HOWITZER ELECTRICAL FIRING TRIGGER
HYDRAULIC TRAVERSING MECHANISM CONTROL HANDLE
TRAVERSING GEAR SHIFTING LEVER
AZIMUTH INDICATOR
FLASHLIGHT

RADIO TRANSMITTER BC-604
ANTENNA BASE AB-15/GR
RADIO SET AN/VRC-3
PATCH CORDS FROM LOADER'S SWITCHBOX
MOUNTING FT-237
LOCKING RECEPTACLE FOR SHELL CASE
GUARD PROVIDED FOR TRANSMITTER
TURRET RADIO TERMINAL BOX
INTERPHONE AND POWER FEED CABLES
RADIO RECEIVER BC-603

The 105 mm howitzer M4 (top right) was ballistically identical with the standard M2A1 field artillery weapon. The breech was rotated 90 degrees with a horizontal sliding breechblock replacing the vertical sliding block of the artillery piece. The same semifixed ammunition was used for both weapons. With this type of ammunition, the projectile is removable from the loosely attached cartridge case, permitting the propelling charge to be adjusted. Removal of powder increments allowed the strength of the propellant to be set over a range from one to seven. The most widely used 105 mm round was the high explosive shell M1. This shell together with the HEAT (high explosive antitank) M67 hollow charge round is illustrated at the right.

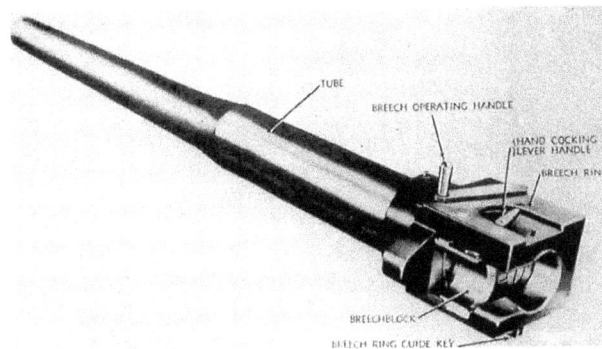

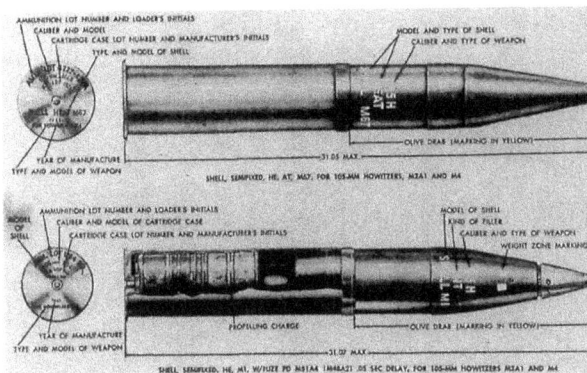

The 105 mm howitzer mount in the three views above and at the right was developed for the medium tank T23. Further work on this mount was discontinued when interest shifted to the T26E1.

The first experimental mounting of the 90 mm gun T15E1 in T26E1 number 1 (above and below) at Aberdeen Proving Ground. The large external equilibrator spring was completely exposed on this early installation. Note the heavy counterweight on the turret bustle necessary to balance the long gun.

Ballistically the 90 mm gun M3 was quite similar to the German 8.8 cm KwK 36, mounted in the Tiger I. Like the 90, it was developed from an antiaircraft gun, in this case the 8.8 cm Flak 18. With the appearance of the high powered German 8.8 cm KwK 43 in the King Tiger or Tiger II tank and its equivalent Pak 43 antitank gun, the need for a higher powered version of the 90 was apparent. To meet this requirement, the 90 mm gun T15 was developed and mounted on a towed carriage. The new weapon was 73 calibers in length and had a much longer, higher capacity chamber. To expedite production of the first two gun tubes, use was made of two forgings already on hand at Watervliet Arsenal. These forgings were slightly undersize for the T15 design and resulted in a gun tube with a slightly smaller chase. These first two guns were designated as 90 mm guns T15E1. A muzzle velocity of 3750 ft/sec was attained with the T30E16 (APCR) type shot. The T33 shot had a muzzle velocity of 3200 ft/sec in the new gun and could penetrate the Panther's frontal armor up to a range of 2600 yards.

The excellent armor protection of the King Tiger and Panther tanks made it increasingly desirable to mount such a high powered weapon in a tank and the most suitable vehicle available was the Pershing, just coming into production. For test purposes, a T15E1 gun was mounted in T26E1 number 1 at Aberdeen Proving Ground and proof fired. These tests indicated it was extremely difficult to handle the long ammunition for the T15 gun inside a cramped tank turret. The complete T33 round was 50 inches in length. Stowage of the long fixed rounds also presented a problem. To correct this, the gun was redesigned to use two piece, separated ammunition. With this system, the projectile was loaded first and the cartridge case was placed behind it. The cartridge case and projectile were stored separately and the nose of the case was plugged. With its chamber redesigned for two piece ammunition, the gun was redesignated 90 mm gun T15E2.

OCM action in March 1945 designated the tank mounting this weapon as the T26E4 and authorized a limited procurement of 1,000 in place of an equal number of M26 tanks. Two temporary pilots were converted by the Wellman Engineering Company and these were readily identified by the external equilibrator springs necessary to compensate for the long heavy gun barrel. Other changes included the installation of a heavier elevating mechanism, turret traveling and turret ring locks, and modification of the cradle to handle the heavier weapon. The turret was balanced by welding a heavy counterweight to rear of the turret bustle and the ammunition stowage was redesigned. The first temporary pilot was shipped to Aberdeen Proving Ground on January 12, 1945 and, following proof firing, was shipped overseas. This first tank was the further modified T26E1 number 1 and was still equipped with the T15E1 gun using the one piece, fixed ammunition. Its adventures in Europe have already been described. The second temporary pilot T26E4 was converted from T26E3 serial number 97 (registration number 30119907) and shipped to Aberdeen for test. This vehicle had the T15E2 gun with separated ammunition.

The second temporary pilot T26E4, serial number 97, under test at Aberdeen Proving Ground. The double external equilibrator springs are now enclosed in protective cylinders. The first temporary pilot was fitted with this same type of equilibrator prior to its shipment to Europe. Since the second pilot was converted from a production T26E3, the turret was fitted with the small oval loader's hatch.

The design for an internal hydropneumatic equilibrator was completed and incorporated in the production vehicles. The new stowage arrangements provided space for 54 rounds of separated 90 mm ammunition. Except for the changes necessary to handle the heavier gun, the T119 combination gun mount was identical with that on the M26. The same coaxial .30 caliber machine gun was carried, but an M71E4 telescope was installed for use with the higher velocity gun. The turret was equipped with a hydraulic power traverse system. Elevation ranged from -10 to +20 degrees and no stabilizer was fitted. With the end of the war in Europe, the number of T26E4s was reduced to 25, all produced at the Fisher Tank Arsenal. Tests at Aberdeen Proving Ground ran through January 1947. Because of the loading problems associated with the two piece ammunition, interest in this type of weapon rapidly faded with the cessation of hostilities and the appearance of better designed one piece ammunition for tank guns. Some of the T26E4s were eventually used as target vehicles.

The T26E4 production pilot converted from T26E3 serial number 84. The hydropneumatic equilibrator has been installed eliminating the external springs.

These views of the production pilot (above) show the new traveling lock fitted for the long barreled high velocity gun.

A production T26E4 was shipped to Fort Knox for tests by the Armored Board. This tank (below) was registration number 30128151, serial number 1405. The necessity for the new traveling lock for the gun is apparent (below left). Unless the gun was carried in an elevated position, the long barrel might strike the ground when crossing rough terrain. The extremes of elevation and depression obtainable with this mount are shown at the lower right. Like other late production Pershings, the T26E4 is equipped with the T80E1 double pin track and the lugs for the power unit hoisting device have been omitted from the turret. The drivers' auxiliary periscopes have also been eliminated.

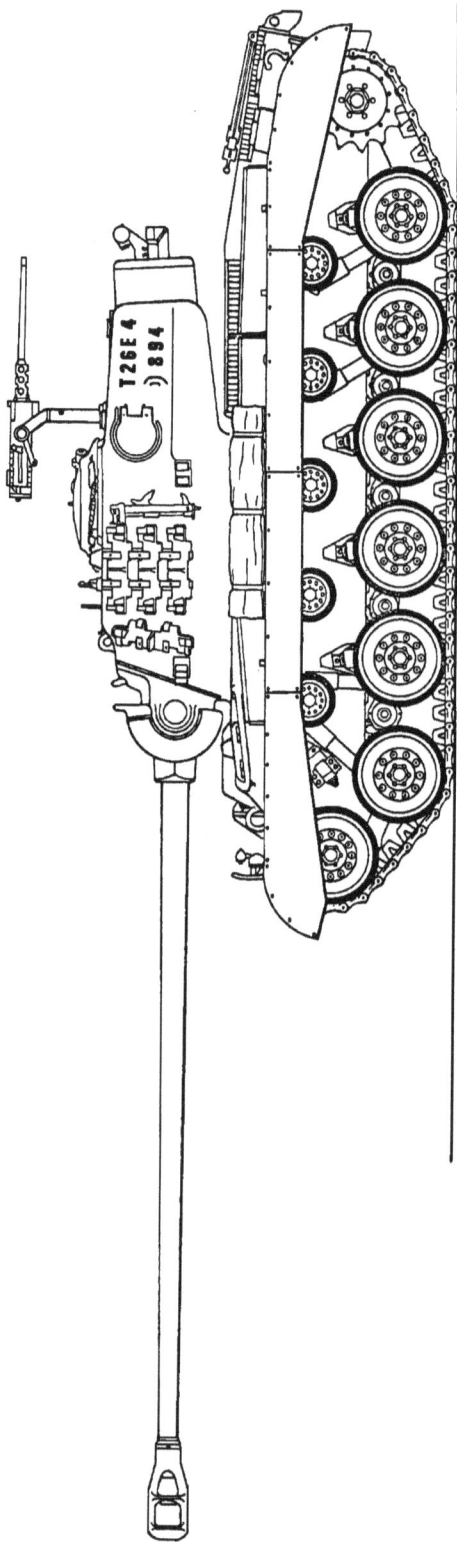

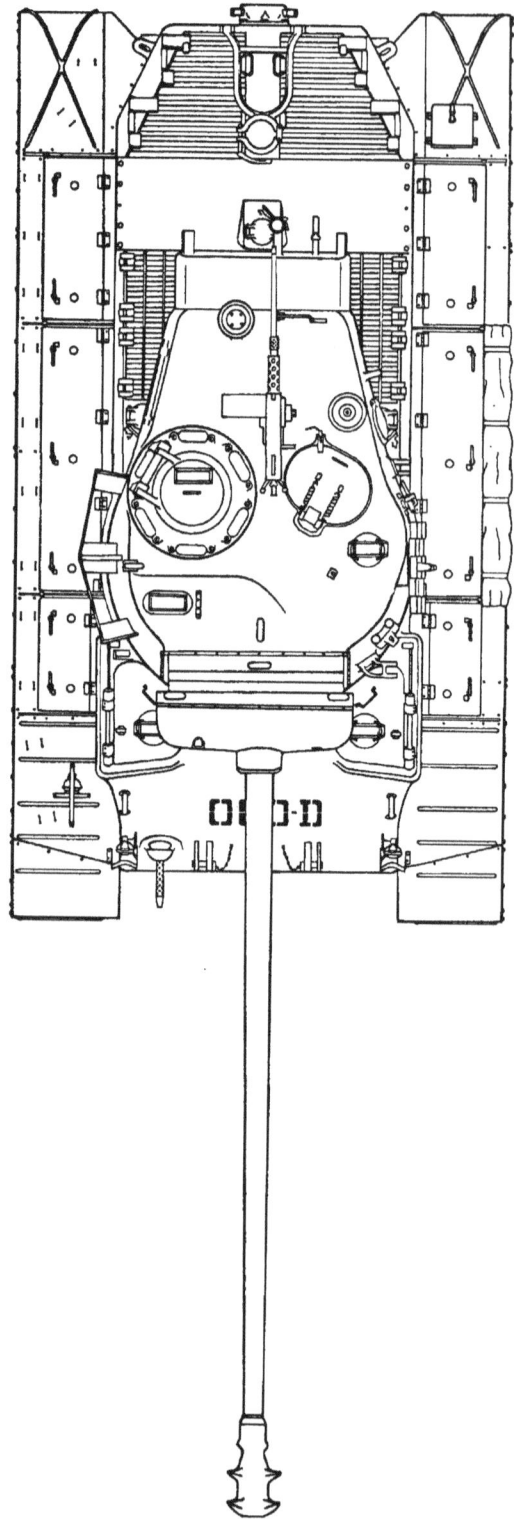

Medium Tank T26E4

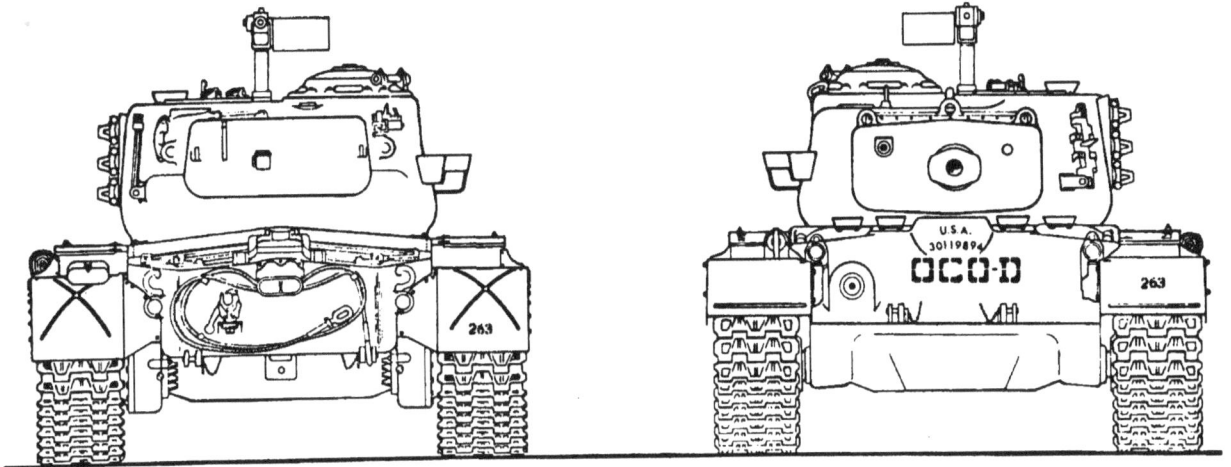

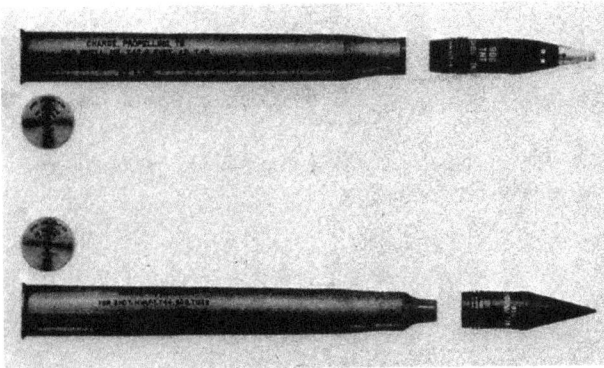

The 90 mm gun T15E2 is shown here with two types of its separated ammunition.

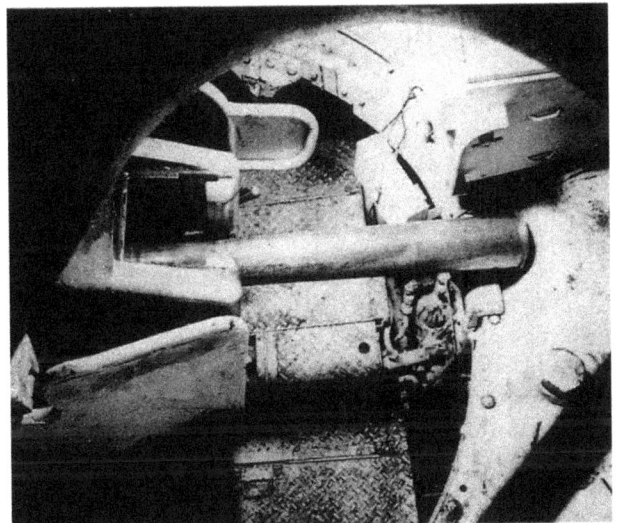

Even with two piece ammunition, the extreme length of the cartridge case caused loading problems. The interference between the case and the turret ring shown above occurred when the gun was in an elevated position.

A T26E4 operating in wooded terrain at Aberdeen Proving Ground. Maneuvering in such areas required extreme care with the long barreled cannon. One mistake would wrap it around a tree. Note the twin .50 caliber antiaircraft machine guns on this tank.

The T26E5 at Aberdeen Proving Ground (above and below) in July 1945. This tank was registration number 30150824 with serial number 10007. Compare the massive turret and front hull castings with the standard Pershing.

The successful combat use of the heavily armored assault tank M4A3E2 pointed out the desirability of an equivalent version of the M26. Preliminary studies produced a design with a sloped front hull having a maximum actual thickness of 4¾ inches. A new turret casting provided an 8 inch thickness. This was intended to give a frontal armor basis of 8 inches for the entire vehicle. Other necessary changes were increased equilibrator capacity to offset the gun shield weight and a lower gear reduction to handle the overall increased vehicle weight. Five inch extended end connectors were used on the T80E1 tracks to reduce the ground pressure.

OCM action on January 18, 1945 recommended the procurement of ten tanks, T26E3, with these modifications and that the modified vehicle be designated T26E5. This was approved on February 8, 1945. However, on March 29 the Ordnance Committee recommended additional increases in the frontal armor and changed the quantity of tanks to 27. Under the new specification, the hull casting would have a thickness of 6 inches at 46 degrees on the upper slope and 4 inches at 54 degrees on the lower slope. The turret casting had a thickness of 7½ inches on the front, 3½ inches on the sides and 5 inches on the rear to balance the heavy gun shield. The new gun shield casting had an actual thickness of 11 inches at 0 degrees. This heavy shield required increased equilibrator capacity plus a 60 pound weight on the recoil guard to balance the gun. Additional thickness was added to the front turret ring splash guard to protect the thin machined surface of the turret adjacent to the ring. It was necessary to decrease the width in the rear portion of the hull escape doors to provide clearance between the doors and the turret. With these changes, the vehicle weight increased to 102,300 pounds, but by use of the 23 inch T80E1 track and 5 inch extended end connectors, the unit ground pressure was held to 11.9 pounds per square inch. The weight increase also required a change in the final drive gear ratio from 1:3.95 to 1:4.47.

146

Production started at Chrysler in June 1945 and the first pilot vehicle (registration number 30150824) was delivered to Aberdeen Proving Ground in July. The tests at Aberdeen indicated that the T26E5 compared favorably in performance with the M26. However, it was necessary to drastically reduce speed over rough terrain to avoid damaging the suspension system. With the end of the war, no additional vehicles were produced and the 27 T26E5s were used for test purposes.

To lower the ground pressure, the track width was increased by replacing the standard end connectors on the double pin tracks with the 5 inch "duckbill" extended connectors. As shown below, external stowage arrangements were the same as the standard late production Pershing.

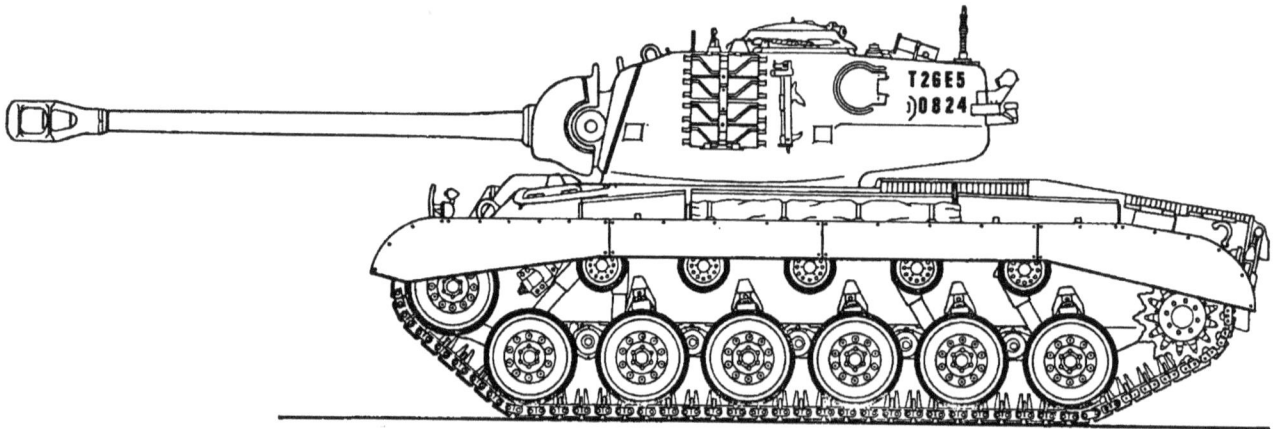

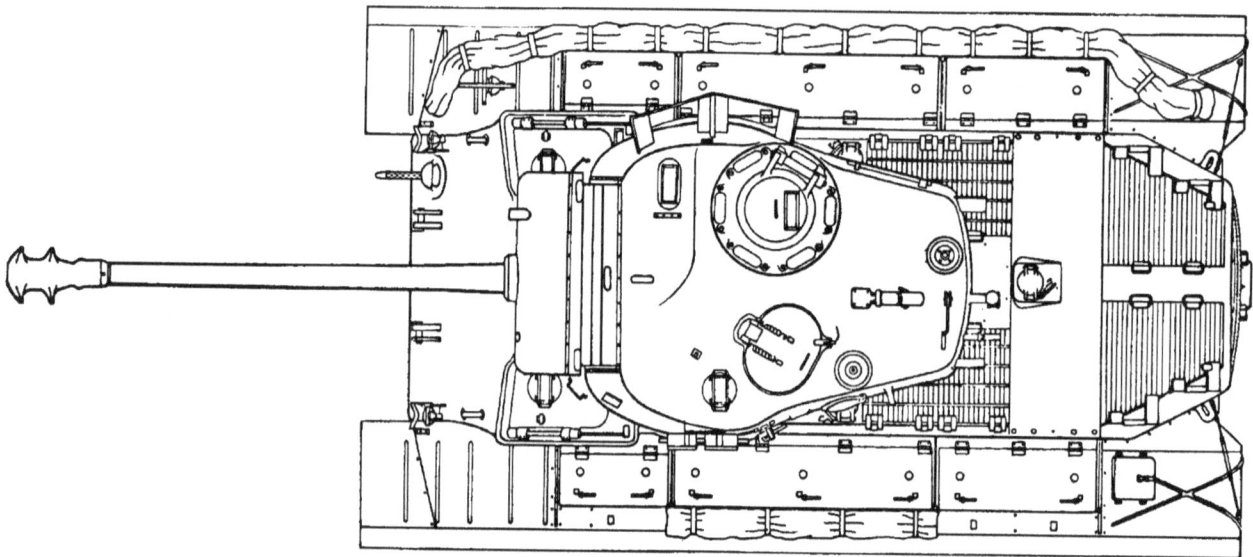

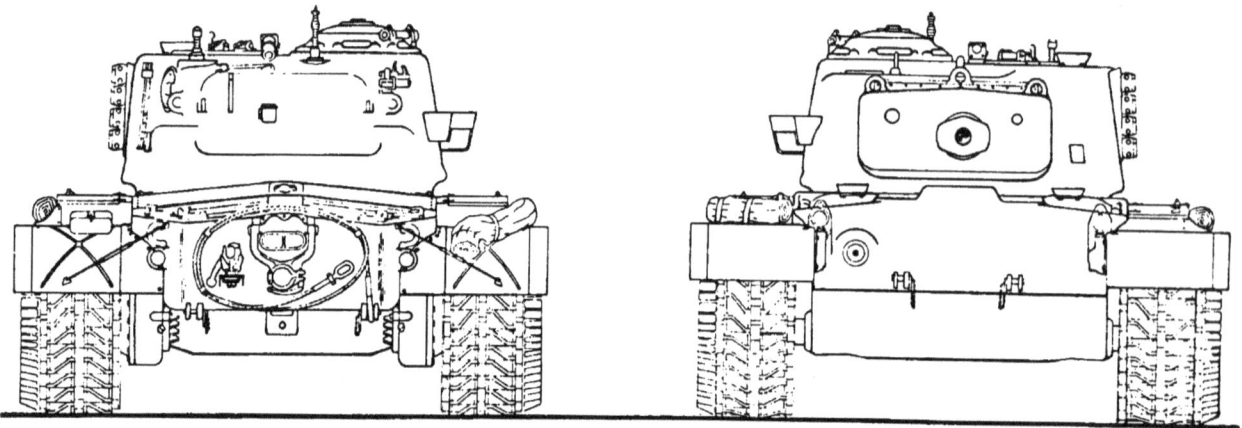

Scale 1:48

Medium Tank T26E5

The third production T26E5, serial number 10009, during tests at Fort Knox. The gun mount cover has been removed and the massive gun shield is exposed. Note an almost square edge has replaced the curved lower part of the standard shield. This was an attempt to minimize the shot trap caused by the lower edge of the shield in much the same way as the Germans modified the late production Panther G. Further details of the turret and shield are shown in the bottom photograph. The numbers chalked on the turret indicate the actual thickness of the casting at that point. Note that the extended end connectors have been removed from the tracks and standard width sandshields fitted. This was necessary to reduce the overall width for transportation.

The second pilot M26E1 (above and below) during tests at Aberdeen Proving Ground.

The M26E1, M26E2, AND M26A1

After the difficulties with separated ammunition in the T26E4, the 90 mm gun T54 was developed to provide a weapon with a fixed, one piece round suitable for use inside a tank turret. The ballistic performance and chamber capacity of the 90 mm gun T15E2 were retained, but the cartridge case was redesigned with shorter and fatter dimensions. This provided a complete round of convenient size to handle and load inside the tank turret. The overall length of the T54 was less than the T15E2, due to its shorter chamber and the muzzle brake was reduced to a single baffle model by machining off the front baffle from a standard 90 mm brake. Except for the muzzle brake, the gun was quite similar in appearance and performance to the German 8.8 cm KwK 43, which armed the King Tiger.

In June 1945, the Ordnance Committee approved the development of two M26 pilots mounting the new gun. These pilots were assigned the designation M26E1. To reduce the congestion inside the turret, a concentric recoil mechanism was developed for use with the T54. Except for the new recoil system, the rest of the gun mount and turret components were basically the same as on the T26E4. The coaxial .30 gun in the T26E4 was replaced by a .50 caliber machine gun HB M2 and the new mount was designated as the 90 mm combination gun mount T126. It also carried an M83C direct sight telescope for the gunner. Ammunition stowage was rearranged to carry 41 of the new stubby fixed rounds. Thirty-six of these were stowed in the floor bins and five were in the ready rack on the loader's side of the turret.

The two tanks were converted by the Martens Ferry Division of Blaw Knox Company and the second pilot was shipped to Aberdeen Proving Ground, while the first was retained at Detroit Arsenal.

The tests at Aberdeen, from February 1947 to January 1949, indicated that the accuracy of fire from the 90 mm gun T54 and the reliability of its concentric recoil mechanism were excellent. The weapon was considered superior to all other U.S. tank guns tested up to that time.

The limited funds available during the immediate postwar period drastically reduced new vehicle development. Following the precedent set prior to World War II, effort was concentrated on developing components. This method had proven successful during the 1930's and had resulted in a suitable line of engine, power train, and suspension components being ready when the need came in 1940. Following this approach, studies were made on improvements in transmission and power plant design.

150

The second pilot was converted from M26 serial number 837 produced by Fisher and is still equipped with the T81 single pin track. The gun travel lock was modified similar to that for the T26E4 in order to hold the long barrel at a higher elevation. The attachment points for the original travel lock are still in place. The long barrel of the .50 caliber coaxial machine gun can be seen protruding from the gun shield.

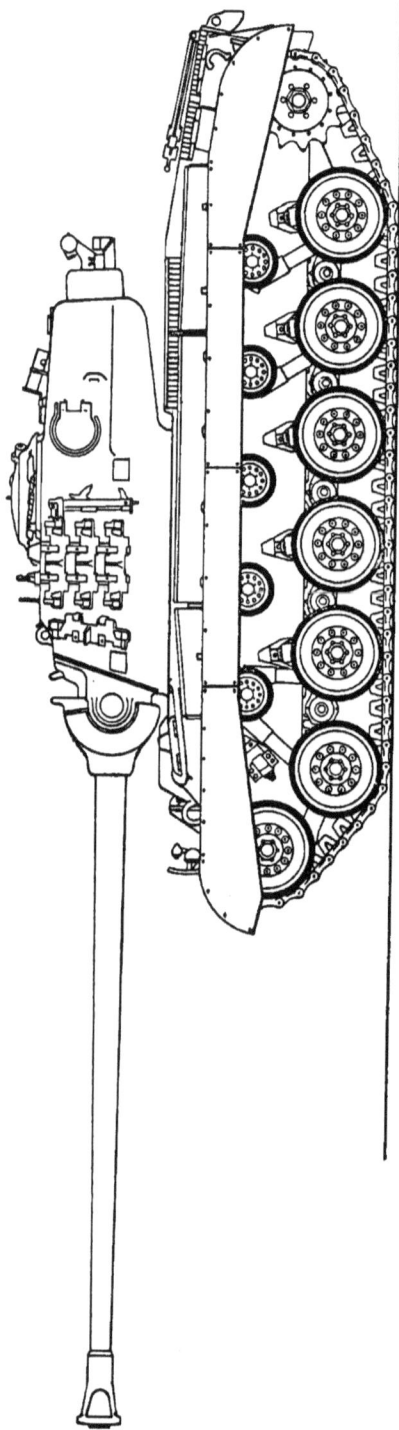

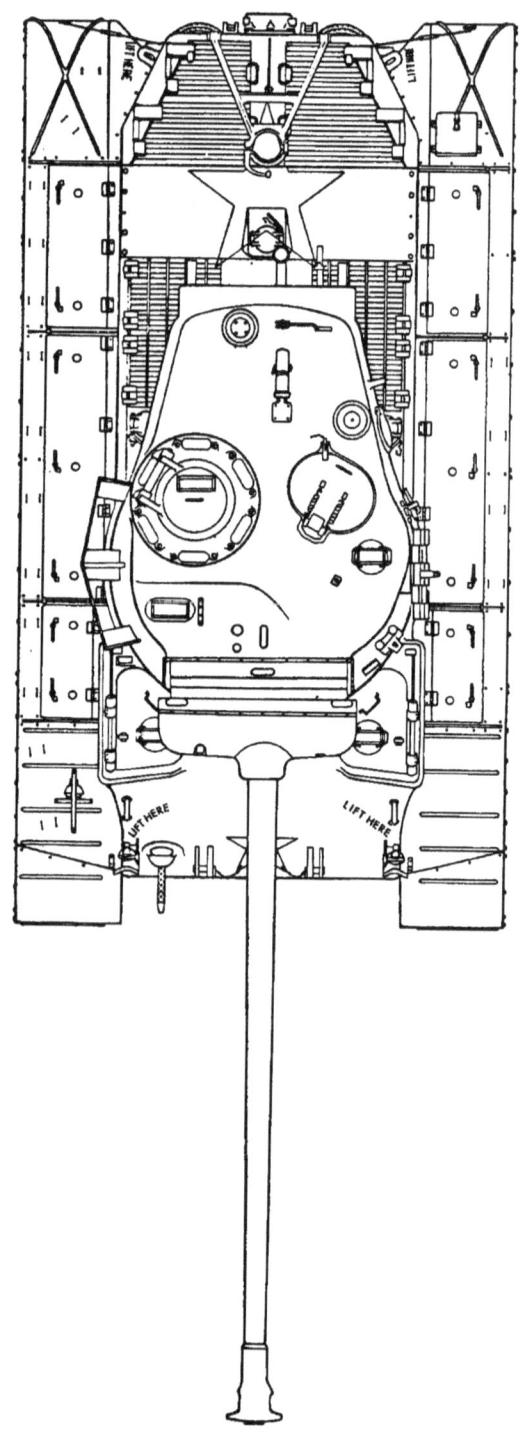

Scale 1:48

Medium Tank M26E1

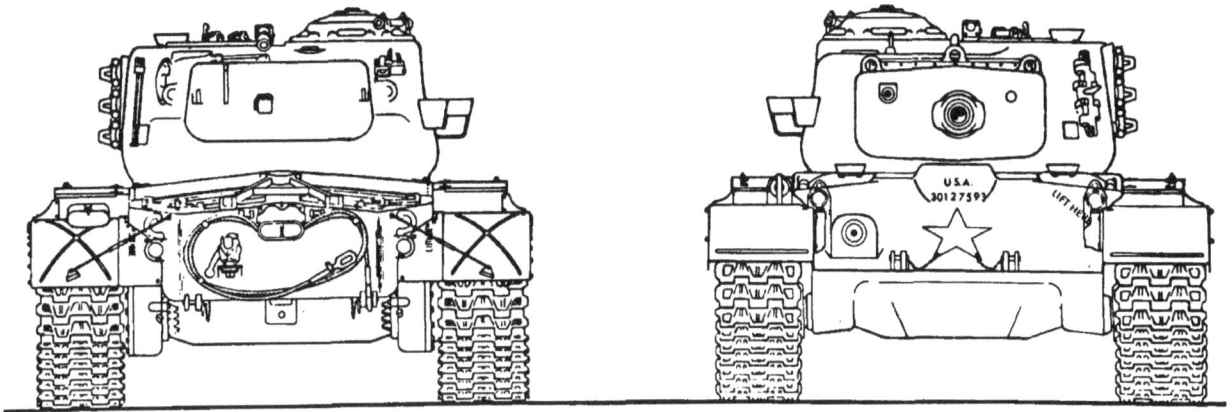

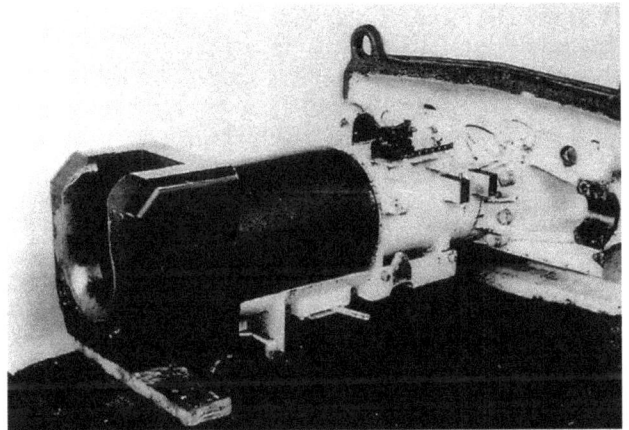

The stubby round for the 90 mm gun T54 is compared (above left) with the standard 90 mm, 76 mm, 75 mm, and 37 mm tank ammunition. Stowage arrangements are shown below in the ready rack (bottom left) and floor bins (bottom right) as viewed from the loader's hatch. The T54 gun in the T126 combination mount (above right) saved considerable space in the turret due to the compact concentric recoil mechanism. The coaxial machine gun and telescopic sight have not been fitted in this photograph.

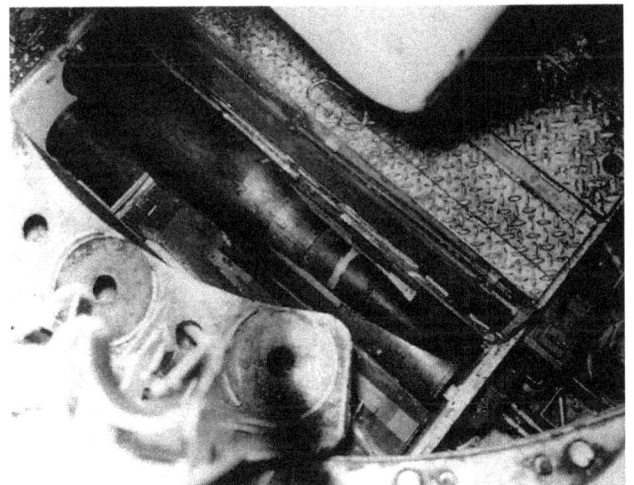

Medium tank M26E2 during the test program at Aberdeen Proving Ground in May 1948. In addition to the rear hull modifications, a major recognition point is provided by the fender mounted mufflers connected to the transverse exhaust pipes.

In 1948, a modernization program was begun to produce an improved version of the medium tank M26. The modernized vehicle was to have a new 810 horsepower air cooled tank engine in combination with a cross-drive transmission.

Both the engine and transmission were products of projects started during the war and continued into the postwar period under the component development program. Together they provided a power unit of 810 horsepower which was over five inches shorter than the 500 horsepower package it replaced. This remarkable achievement was due to the General Motors cross-drive transmission. The Continental AV-1790-3 air-cooled engine was 67.25 inches in length compared to 45.35 inches for the Ford GAF in the M26, but the cross-drive transmission only took up 29.5 inches in length compared to 56 inches for the transmission and differential it replaced.

As its name indicated, the cross-drive was mounted in the transverse direction at the rear of the engine compartment and it combined the functions of the transmission, steering control, and vehicle brakes. As a transmission, it incorporated two hydraulically selected gear ranges driving through a torque converter. A portion of the engine power was transmitted through a mechanical path by-passing the torque converter. This mechanical power, in addition to the power transmitted hydraulically, was applied equally to both output shafts except that in steering, all of the mechanical power was applied to one side to provide the necessary speed difference between the tracks. Braking was accomplished by built-in disc type brakes actuated mechanically from a pedal in the drivers compartment. Steering and shifting were by means of a "joy-stick" type control lever, greatly reducing driver fatigue. Detroit Arsenal installed the new power package in a modified M26 and the vehicle was redesignated as the medium tank M26E2. In May 1948 the M26E2 (registration number 3012420) was shipped to Aberdeen Proving Ground

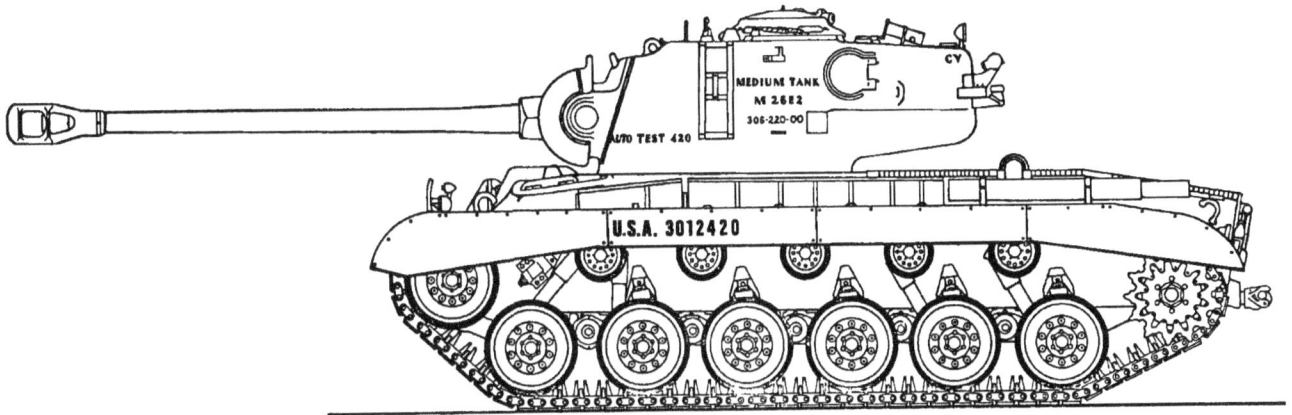

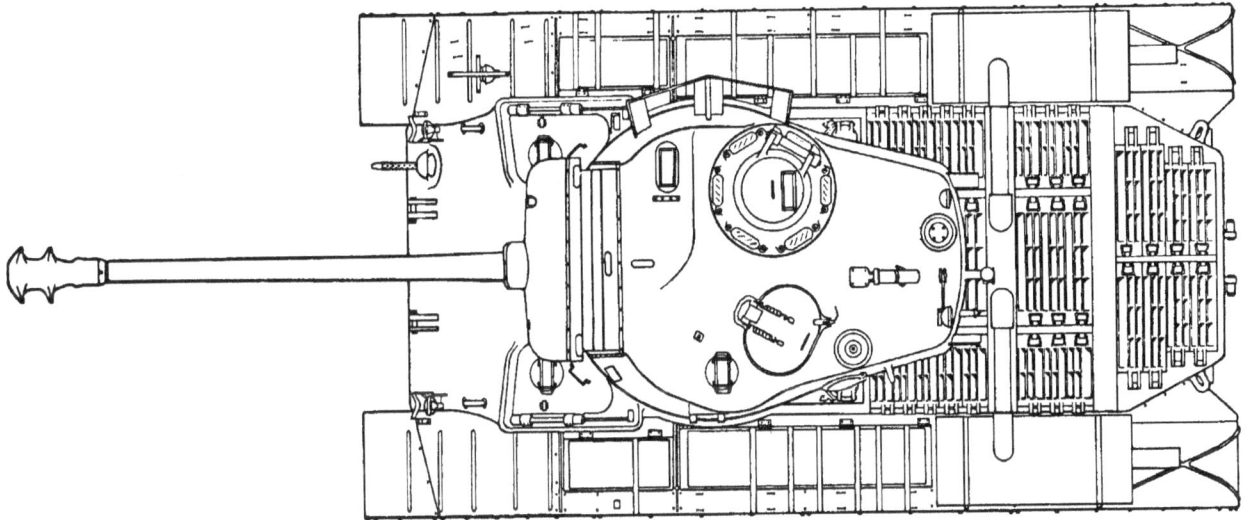

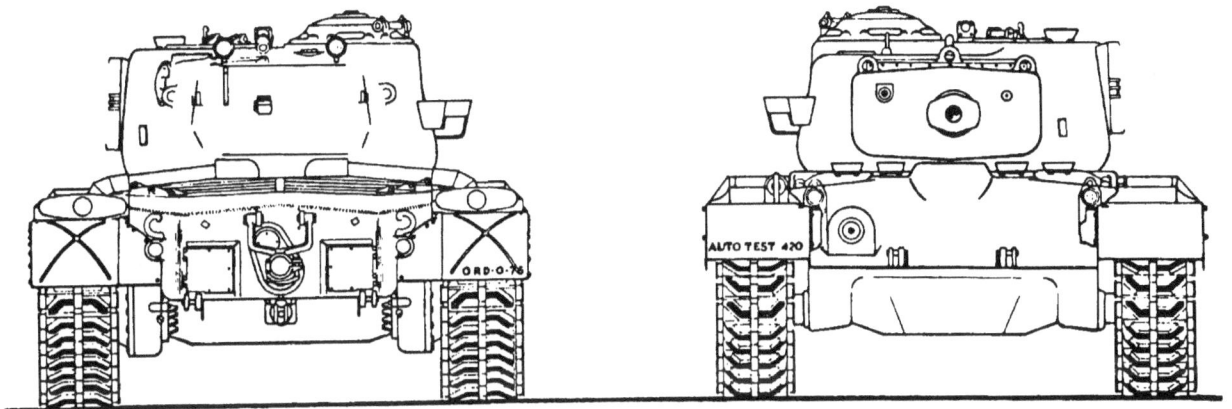

Scale 1:48

Medium Tank M26E2

for test. Like most experimental tanks, it suffered numerous problems, mostly associated with the reliability of the early models of the new power train components. However, the speed and handling characteristics were far superior to the earlier M26 and the test results were incorporated in the newer medium tank T40, later standardized as the M46.

Originally, the M26E2 was intended to carry the high velocity 90 mm gun T54 tested in the M26E1, as well as general improvements in the fire control system. However, the high velocity gun was dropped and efforts were made to improve the standard 90 mm gun M3. A bore evacuator was fitted around the tube just back of the muzzle brake to eliminate powder fumes from the turret when firing. The muzzle brake was redesigned to a lighter weight, single baffle version, since its major function was to direct the blast sideways reducing the obscuration from dust kicked up during firing. With these changes, the gun was installed first on the medium tank T40 and later standardized on the M46 as the 90 mm gun M3A1 in combination mount M73.

When supplies of the new M3A1 gun became available, it was possible to refit some of the standard M26s with the new weapon. For the M3A1, the M67 mount was fitted with a lighter equilibrator spring, because of the smaller torque imposed by the lighter muzzle brake. The gun travel lock was also modified and the designation changed to the combination mount M67A1. Some of these vehicles were also fitted with an elevation stabilizer for the gun. Both the M26A1 and the M46 were destined to serve alongside the standard Pershing in Korea.

The Pershing's offspring, the medium tank T40. Developed from the M26E2, it was later standardized as the M46 and became the first of a series of tanks to carry the name Patton. The fender mounted mufflers are still apparent and the 90 mm gun has been fitted with a bore evacuator. Note that the muzzle brake has been modified to a single baffle model quite similar to that on the M26E1.

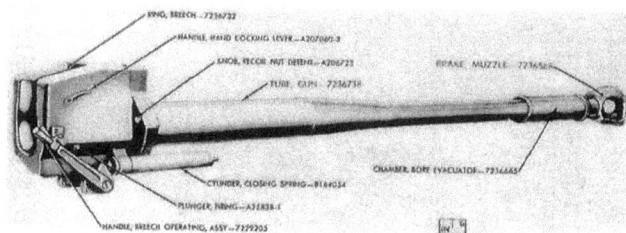

Details of the 90 mm gun M3A1 are shown above and below. This is the weapon that equipped both the M46 and the M26A1.

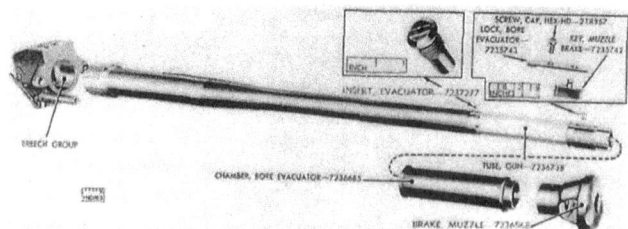

156

An example of the modernized Pershing, the M26A1, on display (above) today in Berlin. It was rescued from the junk pile and restored by the 40th Armor. This tank, serial number 394, originally had the registration number 30127140 and was produced by Fisher in March, 1945. Markings uncovered during restoration, indicated it was used by at least two NATO armies following its U.S. service. Photographs taken at earlier stages of the restoration process are shown at left and below.

An artist's drawing showing the 90 mm antiaircraft gun M1 mounted on the chassis of the T23 medium tank.

VEHICLES BASED ON THE M26

Experience during the early war years underlined the advantage of using the same basic chassis for as many vehicles as possible. A standard set of components greatly reduced the spare parts inventory and simplified maintenance procedures. This lesson gave birth to the "combat team" concept. The "combat team" was a group of vehicles such as artillery motor carriages, cargo carriers, tank recovery vehicles, and similar equipment, all based on a particular tank chassis and using as many standard parts as possible. Following this idea, a variety of vehicles were developed using components from the T20 series tanks.

In early 1943, a design study called for the mounting of the 90 mm antiaircraft gun on the medium tank T23 chassis and in March such an installation was demonstrated to General Devers, General Barr, and other officers. These tests proved useful in the design of the T25 and T26 tanks later in the year.

Following the success of the 155 mm gun motor carriage M12, a lightweight chassis using the engine and power train of the medium tank T23 was proposed as a new artillery motor carriage, designed to carry either the 155 mm gun M1 or the 8 inch howitzer M1. The vehicles were designated respectively, 155 mm gun motor carriage T79 and the 8 inch howitzer motor carriage T80. The cargo carrier T25 based on the same chassis was proposed as an accompanying vehicle. Objections by the Army Ground Forces to the use of the electric drive resulted in the rejection of all three vehicles.

A model of the T23 chassis fitted with the 90 mm antiaircraft gun.

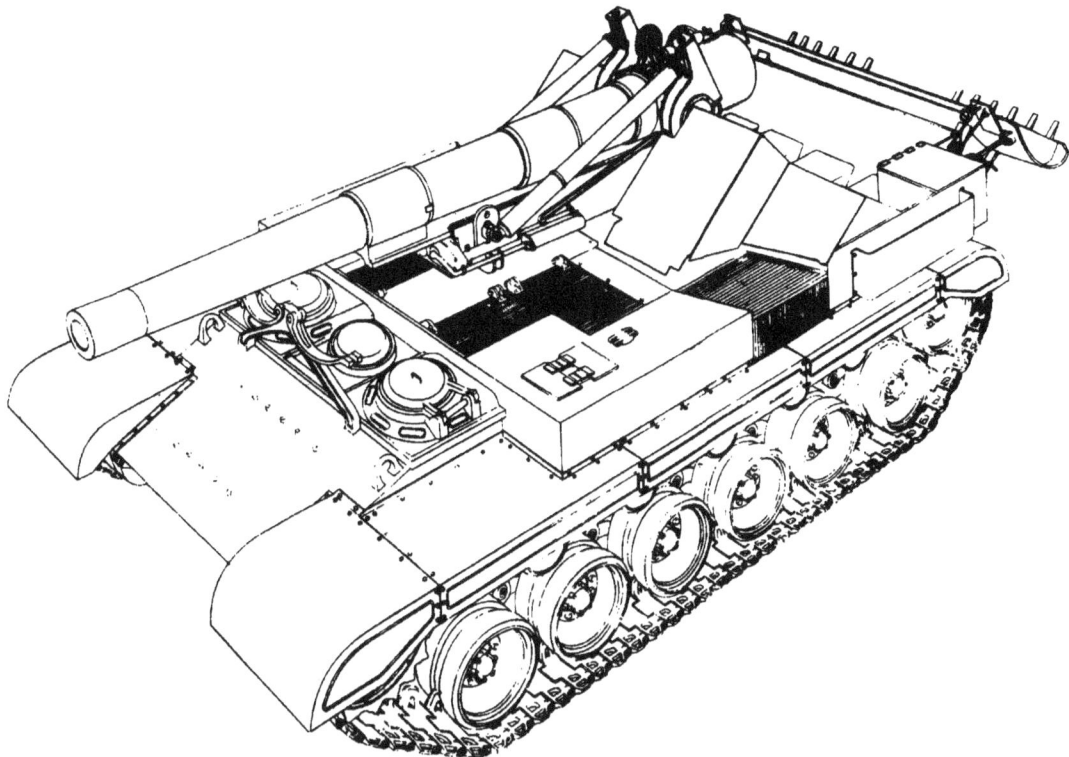

In April 1944, the Army Ground Forces approved the development of the 8 inch howitzer motor carriage T84 based on the medium tank T26E1 chassis. The first pilot (registration number 40215108) was completed at the Chrysler Engineering Division Tank Laboratory and shipped to Aberdeen Proving Ground, arriving in February 1945. Although it used many components of the T26E3, they were drastically rearranged. A new hull was constructed with a maximum armor thickness of one inch and the upper front plate was sloped at 64 degrees from the vertical. Separate, four block, vision cupolas for the driver and assistant driver replaced the customary periscopes. The other six crew members rode at the rear of the carriage. The standard T26E3 power train was rotated 180 degrees placing the differential at the front of the vehicle, the transmission between the drivers, and the Ford GAF engine just behind the driving compartment. The T26E3 suspension, fitted with the 24 inch T81 track, was used with the drive sprockets moved up to the front and the idler and track compensating arm mounted at the rear. By converting to a front drive vehicle, maximum free space was obtained at the rear for mounting the 8 inch howitzer M1. Six emergency rounds of 8 inch ammunition were stowed below the floor at the rear of the carriage.

When received at Aberdeen, the upper front armor was a solid rolled plate carrying two rows of studs for mounting spare track links. Later photographs, taken in April 1945, show that a large access hole was cut in the plate and closed by a bolted removeable cover. No doubt the change was made to permit servicing of the differential without removing the power train from the hull.

The sketch and the photographs show the 8 inch howitzer motor carriage T84 as it was received at Aberdeen Proving Ground.

159

The T84 was a front drive vehicle with the sprocket at the front and the adjustable idler at the rear (above). As received at Aberdeen, the front armor was a solid plate (left) fitted with studs for mounting track links. Compare with the top view (below) showing a removable cover bolted on the front plate. The arrangement of the external stowage is also clearly shown in this photograph.

The 8 inch howitzer motor carriage T84 during firing tests at Aberdeen in April 1945. The shell is being rammed in the left photograph and at the right, the howitzer has just been fired and the tube is in full recoil.

Crew seating arrangement on the motor carriage (right). Six men were carried in the positions shown. The two remaining crew members rode in the drivers' positions in the front hull.

Stowage space for six emergency rounds of 8 inch ammunition was provided in the hull floor behind the howitzer mount (bottom left). The other photograph (bottom right) shows the breech mechanism of the 8 inch howitzer M1 with the block in the closed position. The handwheels for moving the howitzer in azimuth and elevation as well as the sighting equipment can be seen on both sides of the weapon.

Cargo carrier T31 during the postwar period at Aberdeen Proving Ground. The chassis was identical to that of the 8 inch howitzer motor carriage T84 and also was fitted with an access hole in the front armor plate. Note that a steel strip has been welded around the hole to prevent bullet splash from entering under the hole cover.

The cargo carrier T31, based on the same chassis components, was designed as the accompanying vehicle. The carrier used the same front drive arrangement, suspension, and hull construction as the T84, with one inch armor protection on the front and lower sides. The howitzer mount was replaced by a large open top stowage compartment constructed of half inch armor plate. Space was available for 60 complete rounds of 8 inch howitzer ammunition with a special crane for handling the heavy shells. Four of the six man crew rode in the stowage compartment. A .50 caliber machine gun on a ring mount over the assistant driver's hatch provided antiaircraft defense. One pilot T31 was completed by the Chrysler Engineering Division.

The views above and below show the cargo carrier T31 after completion at Chrysler on April 9, 1945. The upper photograph shows the bad weather hoods erected over the drivers' hatches.

The top views (above) show the stowage compartment with and without the canvas cover in place. The racks for the propellant charges are in the center of the compartment. Shell stowage is shown in the left stowage compartment (below left) and the right shell locking tray (below right).

The individual weapons for the crew were located in the small arms stowage on the front bulkhead (below left). Five crew members were issued carbines while the sixth received a .30 caliber rifle M1903A3. The latter was no doubt for use with the antitank rifle grenades shown in right side stowage compartment (below right).

The 240 mm howitzer motor carriage T92 at Aberdeen Proving Ground on 7 July 1945 is shown in the top and bottom photographs. A single post traveling lock was used to minimize the blocking of the driver's vision. This was considered an improvement over the T84.

The T92 was a snug fit on a flatcar even without extended end connectors on the track.

The effective use of the 155 mm GMC M12 in Europe had encouraged Ordnance to consider the development of even heavier self-propelled artillery. In January 1945, self-propelled mounts were proposed for the 8 inch gun M1 and the 240 mm howitzer M1. T26E3 components were used for the engine, power train, and suspension with a lighter weight chassis and the two weapons were designated, the 240 mm howitzer motor carriage T92 and the 8 inch gun motor carriage T93. The chassis were identical in both cases and the weapons could be interchanged. Following the design experience with the 8 inch howitzer motor carriage T84, a lightweight, although longer, hull was constructed with a maximum armor thickness of one inch. Two cupolas, with

three vision blocks each, were provided for the drivers. Unlike the T84, the cupolas were mounted in the sloped front armor, the angle of which was increased to 72 degrees from the vertical. The remaining six members of the eight man crew rode at each side of the gun behind the mount. No ammunition was stowed on either the T92 or T93. Because of the larger and heavier weapons, the chassis was lengthened, requiring an additional roadwheel over the standard T26E3 suspension. Double pin T80E1 tracks were installed and were often fitted with five inch extended end connectors to lower the ground pressure. The power train was reversed, as on the T84, converting the carriage to a front drive vehicle. The cargo carrier T31, with stowage rearranged to handle the heavier ammunition, was intended to be the accompanying vehicle for both weapons.

Original intentions were to ship the pilot carriages to the Pacific Theater for combat testing. It was believed that their powerful weapons would be of great value in reducing the Japanese bunkers and cave defenses. However, the atomic bomb brought about the surrender of Japan before the plan was put into effect.

Five pilot T92s were constructed, but the limited procurement order issued in March 1945 was cancelled by the end of hostilities. Two pilot T93s were built and under test at the end of the war. Once the war was over, interest in the T92 and T93 declined. Tests at Aberdeen revealed the need for a more powerful engine for such heavy vehicles. It was recommended that future heavy self-propelled artillery be designed around the power train components of the new heavy tanks T29, T30, and T32 then under test.

Below: Ready for loading, the 8 inch gun is in firing position with the loading tray in place.

The 8 inch gun motor carriage T93 was identical to the T92 except that the howitzer was replaced by the 8 inch gun M1. Note the extreme overhang in the travel position above, when fitted with the long barreled cannon. The 8 inch gun in firing position at maximum elevation is shown below.

165

This sequence of photographs from side to top clearly shows the general arrangement and external stowage of the T92. Except for the main weapon, the T93 was identical.

Front and rear views of the T92 are shown above. Note the access hole and cover plate on the front armor similar to that on the T84 and T31.

Details of the spade are seen in the view at the left. In this photograph the spade is raised only about halfway to the travel position.

The T92 climbing over the tires onto a tank transporter. Note that the 5 inch extended end connectors have been fitted.

The 240 mm howitzer during firing at Fort Bragg, North Carolina. According to the markings on the vehicle, this was actually the 8 inch gun motor carriage T93 number 1 with the gun replaced by a howitzer.

The T92 at speed during field tests at Aberdeen. The 5 inch end connectors are fitted to reduce the ground pressure.

Construction details which apply equally to the T92 or the T93. The arrangement of the seven road wheels with shock absorbers mounted on the first two and last two is clearly visible (above). Raising the number of road wheels to seven also required the addition of a track return roller giving a total of six per side. The gun mount for the 240 mm howitzer or the 8 inch gun is seen at the left.

A rear view of the chassis (bottom left) shows its appearance before fitting the gun mount.

With the top plate carrying the vision cupolas removed, the entire drivers' compartment is exposed (bottom right). Similar to the tanks, the instrument panel is mounted in the center and duplicate controls are provided for the driver and assistant driver.

Tank recovery vehicle T12 with the boom extended over the front (above left) and left side (above and below) of the vehicle. The front spade is lowered to brace the vehicle when lifting a heavy load over the front.

In 1944, development was started on a tank retriever based on the Pershing chassis. The new design was designated the tank recovery vehicle T12 and was intended to be a service unit for handling heavy turrets, power units, and other components, in addition to its function of removing tanks from the battle area. The standard turret of the M26 was replaced by a lighter turret 1-1/4 inches thick, fitted with a telescopic boom and winch. The turret and boom could be traversed 360 degrees and operated in any position. With the boom at the rear in traveling position, its lifting capacity was 25 tons. Twenty tons could be lifted with the boom at extreme extension, but with the load not more than eight feet, four inches from the vehicle. The boom could be raised or lowered with a load not exceeding 12 tons and the turret could be traversed with a maximum load of eight tons. Two winches, one for the boom and one for line operations, were mounted in the bulge at the rear of the turret. The chassis, hull armor plate, and controls were the same as the M26 tank. T80E1 double pin tracks with five inch extended end connectors gave a total track width of 28 inches. Normally, a crew of six was required. The .30 caliber machine gun was retained in the bow mount and a .50 caliber antiaircraft machine gun was ring mounted at the right on top of the turret. A vision cupola was provided for the vehicle commander on the left side of the turret roof.

Although cancelled by the end of the war, a design was proposed for the mine resistant vehicle T15E3 based on the M26 tank. This vehicle was to be fitted with heavy belly armor and mine resistant tracks to permit greater safety when used as a prime mover for mechanical mine exploding devices.

The top view (at right) shows the boom in the traveling position over the rear. The open, ring type, machine gun mount and the commander's cupola are visible on top of the turret.

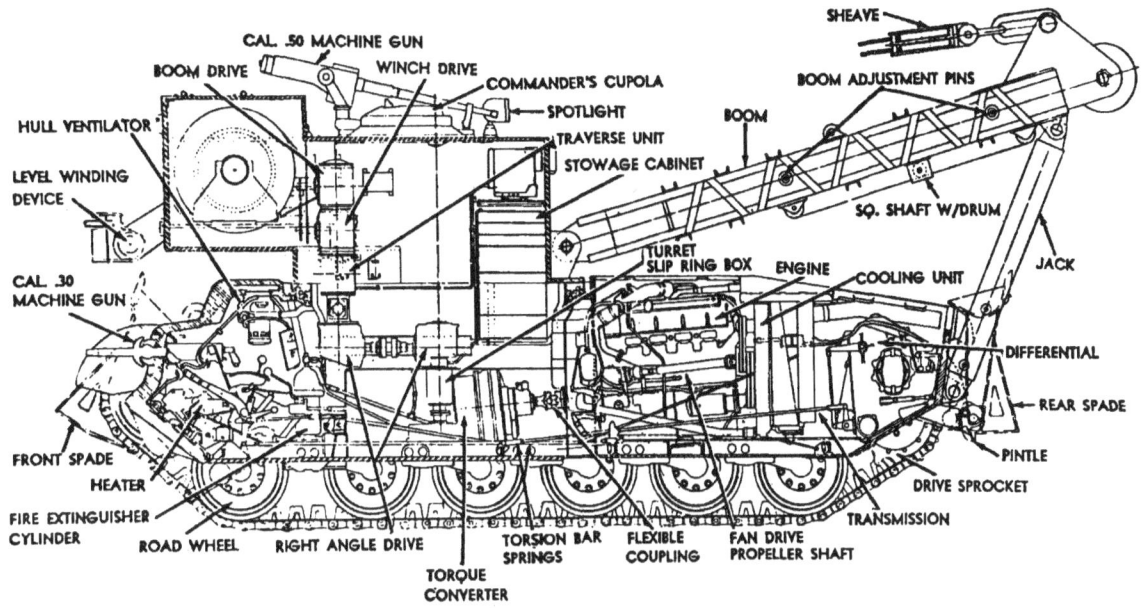

The interior arrangement is shown in the sectional view above. Details of the spades in the raised and lowered positions can be seen below.

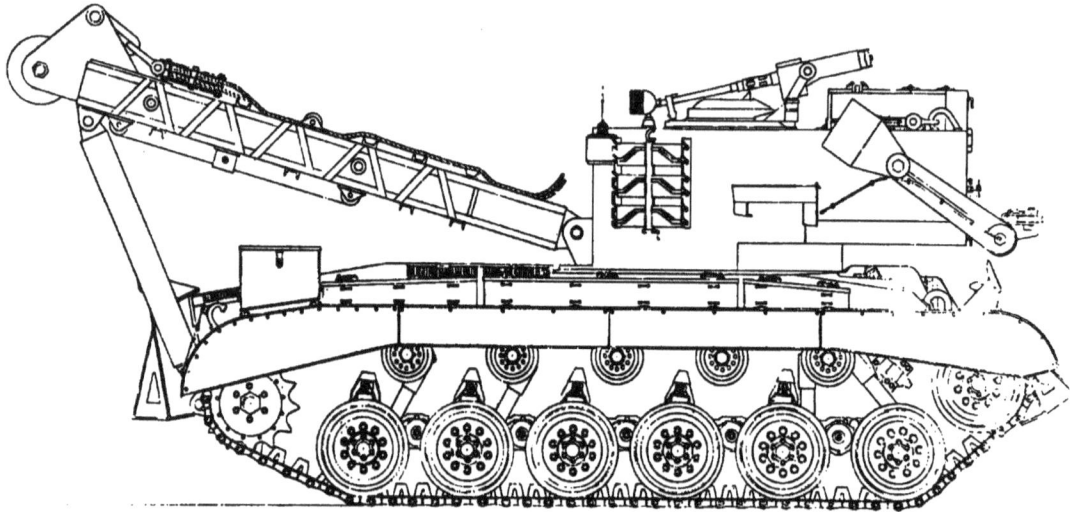

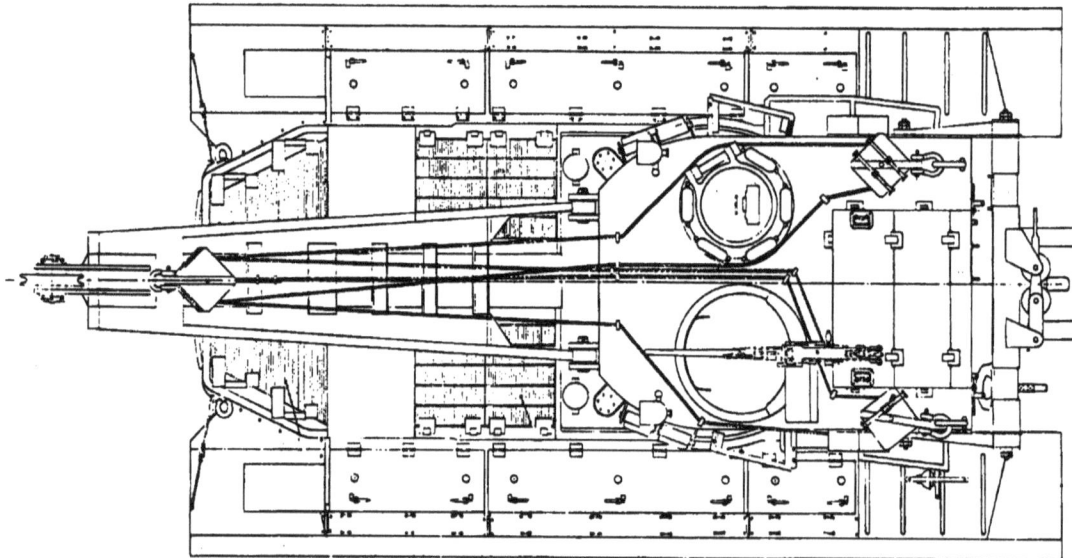

Scale 1:48

Tank Recovery Vehicle T12

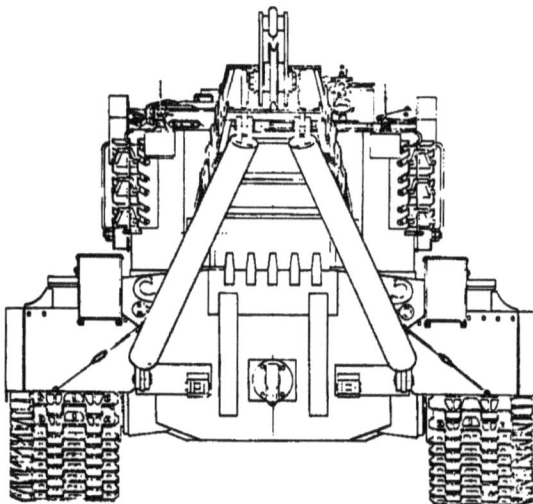

Unlike the other drawings in this book, these views of the T12 were reproduced directly from the original Arsenal drawings. An error is noted in the rear view. The tank is shown equipped with T81 single pin tracks fitted with extended end connectors, an obviously impossible condition. No doubt a wartime draftsman tried to save some time by tracing the tracks from a Pershing drawing and adding the "duckbill" connectors. For a proper view of the T80E1 double pin track with the extended end connectors, see the drawing of the T26E5 on page 148.

PART III

FINAL SERVICE

The Soviet built T34/85 medium tank.

KOREA

Before dawn on Sunday, June 25, 1950, units of the North Korea People's Army (NKPA), attacked southward across the 38th parallel. The attack force numbered about 90,000 men and included seven infantry divisions and one armored brigade. The invasion was without warning and achieved complete surprise, although many ominous signs during previous weeks had indicated the possibility of an impending attack.

On June 8, 1950, newspapers in the northern capital of P'yongyang published a manifesto calling for the election of a national parliament from both north and south to meet in Seoul on August 15, the fifth anniversary of Korea's liberation from Japan. Apparently the northern leadership believed that a "Blitzkrieg" invasion would be able to destroy the Republic of Korea (ROK) Army in the south in a single rapid thrust. The elections could then be held under northern control in the occupied area and a single communist government established in Seoul by the middle of August. To achieve such a rapid and complete victory, a striking force of great power and mobility was required. This was provided by the 150 Soviet T34/85 tanks available to the NKPA. The 105th Armored Brigade was equipped with 120 T34s and the remaining 30 were assigned to the 7th Infantry Division at Inje in east central Korea just prior to the attack. The tanks of the 105th led the drive to the Han river and, after the capture of Seoul, the unit was redesignated the 105th Armored Division.

The T34 which performed so effectively in these operations was the same vehicle widely used by the Soviet Army during the latter part of World War II. The combination of the 85 mm gun and the well armored, highly mobile chassis was a formidable fighting vehicle, difficult to defeat even with the best of antitank weapons. In June 1950, such weapons were not available to the South Korean Army. They had only the early World War II 37 mm antitank guns and their artillery was limited to a few U.S. 105mm M3 howitzers. This was a short barreled version of the standard M2A1 weapon, with a lower muzzle velocity. Little or no HEAT ammunition was provided for antitank defense. The infantry units were equipped with the obsolete 2.36 inch "Bazooka" rocket launcher left over from World War II and its ammunition was old and unreliable.

In June 1950, the South Korean Army had no tanks and the U.S. forces in the Far East were not much better off. Each of the four U.S. divisions on occupation duty in Japan was assigned a tank battalion, but each had only one company (A) organized and equipped. Even though some were designated as heavy tank battalions, the only equipment provided was the M24 light tank. This vehicle was satisfactory for occupation duty and minimized the damage to light bridges and roads in Japan.

175

A U.S. M24 light tank moves through a Korean village enroute to the front on July 8, 1950.

Despite their lack of suitable weapons, the South Koreans fiercely resisted the initial advance of the T34s. Antitank weapons were improvised using satchel charges and hand grenades. South Korean soldiers even climbed on top of some tanks trying to pry open a hatch to throw in a hand grenade. A few T34s were destroyed, but the usual result was that the attackers were killed and the tanks rolled on unharmed. The appearance of invincibility had its effect on morale and it rapidly became difficult to find volunteers willing to attack the tanks.

With the entry of American ground troops into action, the M24 light tanks were moved from Japan to Korea. They were first committed to battle July 10, 1950 near Chonui. It was rapidly apparent that the M24, although a good reconnaissance vehicle, was no match for the T34 and two were lost during the afternoon of the 10th.

Dug in to protect its light armor, an M24 awaits the appearance of the enemy, July 9, 1950.

The M24 light tanks engage the enemy near Chonui on July 10, 1950.

The original Signal Corps caption indicates the M24 shown in the two views at the right was the first U.S. tank in action in Korea on July 10, 1950. The crew is shown cleaning up the tank, nicknamed "Rebels Roost", after returning from the front.

177

One of Lieutenant Fowler's Pershings test firing its guns near Taegu on July 22, 1950, six days before moving to Chinju.

Immediately after the outbreak of war, the Ordnance stores in Japan were ransacked to provide all available weapons to the forces ordered to Korea. On June 28, three Pershing tanks were located at the Tokyo Ordnance Depot. All three were in very poor condition requiring major repair. This work began immediately and was completed on July 13th. The tanks, manned by 14 enlisted men under the command of 1st Lieutenant Samuel R. Fowler, arrived at Pusan on July 16. Lieutenant Fowler and his crews had been trained on the M24 light tank so they immediately began to zero their guns and familiarize themsleves with the heavier M26. It soon became obvious that the rebuilt engines left a lot to be desired. The standard fan belts had not been available in Japan and the substitutes stretched after a short period of operation allowing the fan to stop and the engine to overheat. New belts were requested, but had not arrived when the three tanks were shipped by rail to Chinju arriving early in the morning of July 28. As the only medium tanks in Korea, they had to be used to try and stop the NKPA drive toward Pusan from the southwest. Fowler and his crews, still awaiting their fan belts, remained with the tanks near the rail depot south of the Nam River until the morning of July 31, when enemy troops entered Chinju. Knowing his tanks would not withstand a long road march, Fowler telephoned Masan and was told that railway flatcars were on their way to Chinju to evacuate the Pershings. He then decided to wait,

even though the infantry was pulling back from Chinju. About 1300 hours, an enemy patrol of platoon strength approached the rail station and was taken under fire by the three tanks. The .30 and .50 caliber machine guns wiped out the enemy force, but Lieutenant Fowler was wounded during the exchange. His crew placed him in his tank and, as other enemy troops were in the vicinity, the three Pershings moved eastward along the road toward Masan. After about two miles, they were halted by a blown bridge and dismounted to proceed on foot. A litter was being prepared for Fowler when an enemy ambush opened fire, forcing some of the men to seek shelter under the bridge. Master Sergeant Bryant Shrader was still inside one of the tanks and he returned the fire and maneuvered the Pershing to pick up six other men through the bottom escape hatch. Unable to reach the men under the bridge, he drove back down the road to Chinju, stopping at the Nam River. Here the overheated engine stalled and would not start again and the seven men abandoned the tank. Hiding from enemy troops, Shrader and his men managed to work their way back to the American lines at Masan, but Lieutenant Fowler and the men left at the bridge were killed or captured. Eight men and the only three medium tanks in Korea were lost mainly through the lack of such insignificant components as fan belts. It would be difficult to provide a better modern example of the loss of a battle "for the want of a horseshoe nail".

178

Two additional views of Lieutenant Fowler's Pershings while zeroing their guns at Taegu. Since the crews had trained on the M24 light tanks, they were forced to learn the operation of the Pershing on a crash basis.

USA. 30127945

A Pershing on the road to Sach-on in action against a North Korean thrust in the Masan-Chinju area on August 10, 1950.

The key role played by the Soviet built tanks made it essential that U.S. medium tanks swiftly be transferred to Korea. The Army notified General MacArthur July 10th that it planned to ship three battalions of medium tanks directly to the Far East as the quickest means of getting these vehicles with trained crews into action. The battalions selected

Troops of the 1st Cavalry Division are introduced to the newly arrived Pershing before returning to the front on August 10, 1950.

were the 6th equipped with M46s, the 70th with M26s and M4A3s, and the 73rd equipped with the M26. So once again, the Pershing and the Sherman went to war, unloading at Pusan, Korea on August 8th.

While arrangements were being made to ship the battalions from the United States, Eighth Army activated the 8072nd Medium Tank Battalion in Japan. This unit was equipped with World War II Shermans salvaged from the Pacific islands and repaired in Japan. Company A landed at Pusan on July 31st followed by the remainder of the battalion on August 4th. Three days later the unit was redesignated as the 89th Tank Battalion.

The new battalions were immediately deployed to strengthen the defense of the perimeter protecting Pusan. The 6th was moved to the Taegu area as part of the Eighth Army reserve, while the 70th was sent to support the 1st Cavalry Division. The 73rd was split up with A Company moving to Ulsan, B Company going to the Kyongju and Kigye area, and C Company reinforcing the 27th Infantry north of Taegu in the region soon to be known as the "Bowling Alley". Additional reinforcements, including the 72nd Tank Battalion and the 1st Marine Tank Battalion, also arrived bringing the tank strength in the Pusan perimeter to over 500 by late August. Except for the 6th equipped with M46 Pattons, the medium tanks consisted of approximately equal numbers of Pershings and Shermans.

Above: Pershings assigned to the 89th Tank Battalion firing on North Korean positions in the area of the U.S. 29th Infantry Regiment on August 25, 1950. The earth ramps were used to increase the elevation and obtain the maximum range with the 90.

Right: Four M26s firing on a North Korean observation post across the Naktong River during the fight for the Pusan perimeter.

Below: M26 from C Company, 72nd Tank Battalion in action along the Naktong River line on August 31st. In all of these photographs the tanks are being used as mobile artillery.

The first encounter between the Pershing and the Soviet T34 occurred just before dark on August 17th during the first battle of the Naktong Bulge. The latter name was given to the loop of the Naktong river just west of Yongsan, where elements of the North Korean 4th Division had crossed and penetrated several miles east of the river. The U.S. 24th Infantry Division, including the 1st Marine Provisional Brigade, counterattacked to destroy the enemy forces east of the Naktong. On the evening of the 17th, troops of the 1st Battalion, 5th Marines were digging in after taking hill 102. This position overlooked the road crossing the pass between Cloverleaf hill on the north and Obong-ni ridge to the south. It was along this road that the marines observed three T34s, followed later by a fourth, moving eastward toward the pass. Bazooka and recoilless rifle teams were positioned to ambush the enemy tanks and three Pershings were called forward to positions covering the eastern end of the pass. As the first T34 rounded a bend in the road it was hit by both recoilless rifle and bazooka fire which stopped the vehicle, but did not silence its guns. The first Pershing now took a hand, setting it on fire with a single 90 mm round. Small arms fire killed one crew member who attempted to escape. The bazooka teams accounted for the second T34 the moment it came into sight and as the third made its appearance, two Pershings destroyed it with 90 mm fire. The fourth T34 had dropped well behind the others and was destroyed by air action.

Above: The Marine Corps Pershing shown here on August 17, 1950 is in the area where the first action occurred with the Soviet T34. Note the empty 90 mm cartridge cases along the side of the road.

Below: Another Marine Corps M26 photographed in the same area two days earlier while supporting 1st Cavalry Division troops on August 15th.

Above: Another Marine Corps M26 near Yongsan. Note that all of these Marine Corps tanks are still equipped with the original T81 single pin tracks.
Below: A Pershing from Company C, 73rd Tank Battalion moves up to support the 27th Infantry in the Taegu-Waegwan sector on August 20, 1950.

The "Bowling Alley" looking north toward enemy territory on August 21, 1950. A Soviet built SU-76 self-propelled gun has been destroyed on the road and U.S. artillery fire can be seen falling on the enemy positions up the valley.

Following the fight on August 17th, the Pershings were frequently in action against the enemy armor. On the following day the scene shifted to the Taegu front further north. Here the 27th Infantry attacked along the Taegu-Sangju road to relieve enemy pressure on Taegu. About two miles north of Tabu-dong, enemy resistance held up the advance of the ROK units in the hills on both flanks of the American force. The 27th then halted and formed a perimeter across the road, just north of the village of Soi-ri. The Pershings of Company C, 73rd Tank Battalion were in support, with one platoon in the front lines and four more tanks in reserve just back of the line. That evening, August 18th, they fought off the first of seven NKPA night attacks which were to continue through August 25th. Every night, the enemy attacked south with infantry, tanks, and self-propelled guns following the straight section of road bisecting the American positions. Apparently during the wild action on the night of August 21st, this area was nicknamed the "Bowling Alley" by men of F Company, 27th Infantry. On this night, the enemy T34s were firing straight up the road trying to knock out the Pershings in the U.S. positions. The tracers of the armor piercing projectiles streaking up the road and the thunder of gunfire rolling back from the hills on either side combined to give the impression of a gigantic bowling alley.

Enemy attacks continued through the night of August 24th, by which time confirmed enemy losses included 13 T34 tanks and 5 self-propelled guns. U.S. troops were relieved on the night of August 25th by ROK forces.

A Pershing nicknamed "Pat" from C Company, 73rd Tank Battalion loads ammunition in the "Bowling Alley" on August 21st.

Pershings move between three T34s destroyed in front of the U.S. positions. One T34 is still carrying its jettisonable fuel tanks. These photographs taken on August 24th show the typical cluttered appearance of the tanks in action with all sorts of items stowed in a non-regulation manner. In the bottom photograph, the Pershing is firing at a group of North Koreans laying a mine field further up the valley. The muzzle blast kicks up the dirt in front of the tank.

Above: An M26 bypasses a destroyed bridge near Tabu-dong.
Below: Another Pershing in action in the same area. Both photographs were dated August 24, 1950.

Pershings of the 73rd Tank Battalion move into position on August 27, 1950. The tanks in these views were marked with a girl's name on the front and sides. The top tank also has the crew member's name painted by his station.

In the bottom photograph, the air from the engine cooling fans causes the aircraft recognition panel to billow above the rear deck.

After the early encounters, many other actions occurred between the Pershing and the T34, but the major tank work was in infantry support. Like the German Tiger six years earlier, the Pershing had no difficulty in dealing with the T34 on a tank for tank basis. The 90 mm gun could easily destroy the Soviet built tanks at all normal combat ranges. However, the high power to weight ratio and wide tracks of the T34 gave it superior mobility, particularly in rough country. The Pershing continued to serve widely, particularly during the first six months of the war. However, when the action moved into more mountainous country, the M4A3 with the same engine, but lighter weight was preferred. By the end of the war, the old Sherman had again become the most widely used tank in action.

Above: Men of the 9th Infantry Regiment riding on an M26 move to meet an enemy attack across the Naktong River, September 3, 1950.

Below: An example of field camouflage applied to the Pershing. Another interesting point is the .30 caliber machine gun fitted to the forward pedestal mount. The .50 caliber weapon is retained on the rear mount.

Tanks of the 73rd Battalion wait to board LSTs at Pusan for the move north to Inchon. Note the two versions of traveling gun locks in these photographs. The type shown below attached to the rear plate is the later production version.

Above: An M26 of the 72nd Tank Battalion supporting troops of the 9th Infantry Regiment on the Naktong River line, September 1950.
Below: A Pershing moves into position in the Tabu-dong area while the infantry awaits orders to move up.

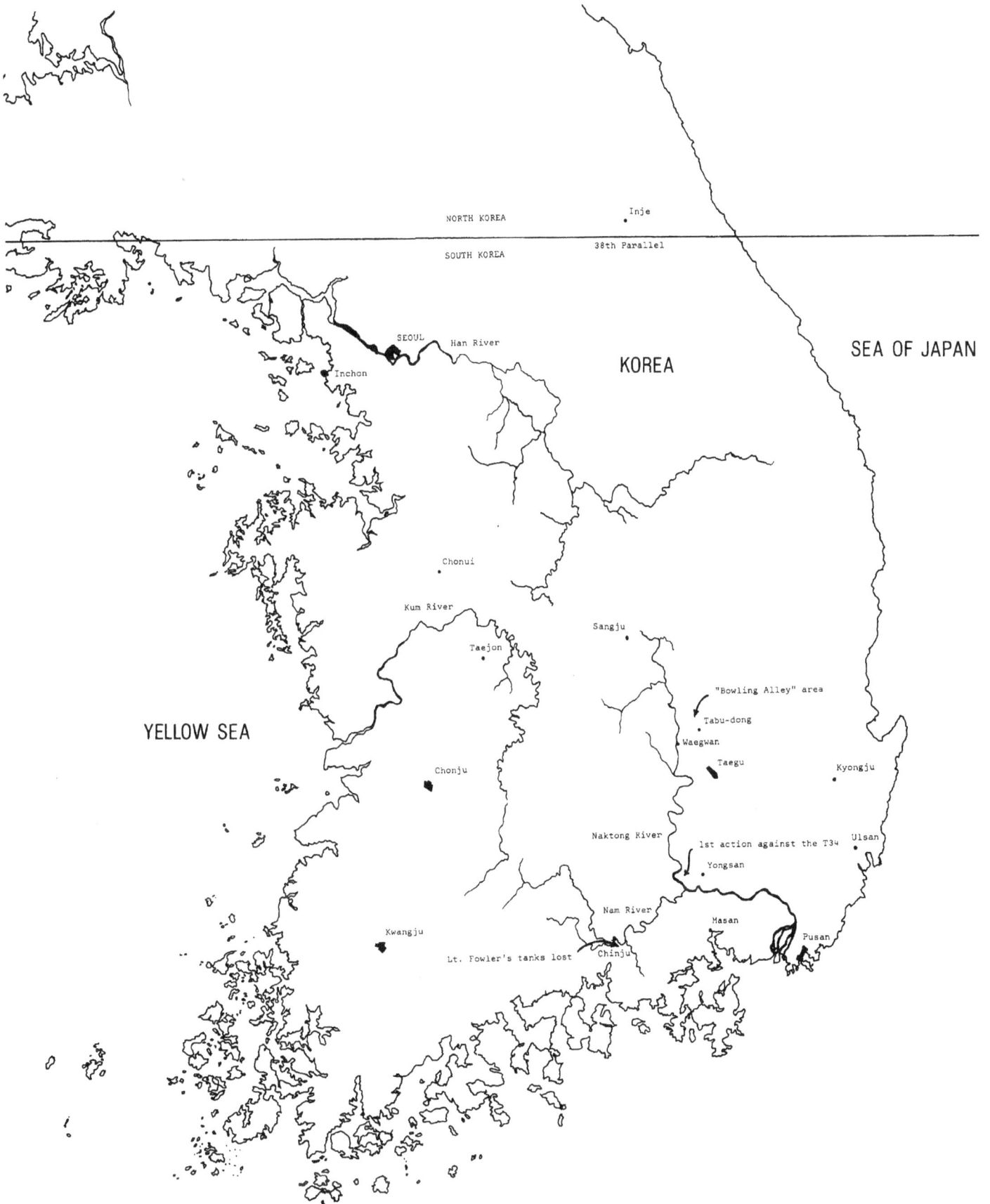

South Korea
Area of Operations, June-December 1950

NORTH KOREA

SOUTH KOREA

Inje

38th Parallel

KOREA

SEA OF JAPAN

SEOUL

Han River

Inchon

YELLOW SEA

Chonui

Kum River

Sangju

Taejon

"Bowling Alley" area

Tabu-dong

Waegwan

Taegu

Kyongju

Chonju

Naktong River

1st action against the T34

Ulsan

Yongsan

Nam River

Masan

Pusan

Kwangju

Chinju

Lt. Fowler's tanks lost

The T34 could also destroy the Pershing. This shows an 85 mm hit on the front of an M46 Patton which had the same frontal armor as the Pershing. The shot struck perfectly at the base of the lug which provided a shot trap giving much the same effect as a vertical plate. This illustrates the danger involved with any attachments that interrupt the smooth shape of the front armor.

An 85 mm hit in the shot trap at the bottom of the Pershing's gun shield. The shield has been removed and the turret rotated to face the side of the tank.

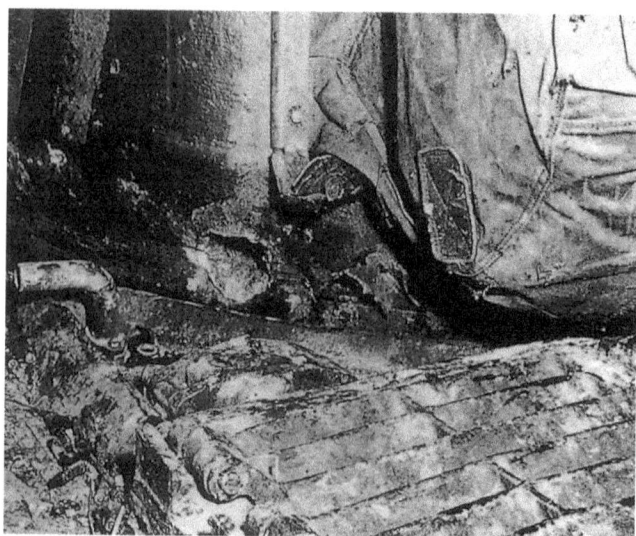

Even the smaller enemy antitank guns could be deadly. The hit shown here on an M46 of the 64th Tank Battalion came from a Soviet built 57 mm gun.

Above: Two M26A1s entering Seoul on September 29, 1950. These vehicles can easily be identified by the bore evacuator on the 90 mm gun and the absence of fender mounted mufflers characteristic of the M46.

Below: An M45 of the 6th Tank Battalion crosses the Kumho River on an underwater crossing of rocks and sand bags reinforcing the river bed, 18 September 1950.

In addition to Korea, the Pershing served widely in Europe with the U.S. and other NATO forces. An M26 of the 63rd Tank Battalion crossing the Main River at Aschaffenburg, Germany during Exercise "Combine" in October 1951.

CONCLUSION

The Pershing story might well be summarized by the words of Captain Elmer Gray replying to the tank crews at Aachen when they asked if the Pershing was equal to the German King Tiger and Panther. His answer was, "Hell no, but it is the best tank we have yet developed and we should have had it a year earlier". The reasons for the delay of the Pershing were many, among them, the commitment of the Army Ground Forces to the exploitation role and the doctrine which specified that tanks should not fight tanks. The latter view, strongly supported by General McNair, retarded the development of heavily armed and armored tanks until experience with the German heavy weights proved that such encounters could not be avoided.

Anticipating the requirements of 1944, General Barnes, in October 1943, had urgently requested immediate production of 500 each of the T23, T25E1, and T26E1. The request was rejected on the grounds that no requirement existed for these tanks. This resulted, in part, from the attitude of the Armored Force which wanted the 90 mm gun, but not in a new tank. The Armored Force believed that a new tank chassis would not be fully developed in time to be of effective use in the war and preferred to have the familiar Sherman modified to mount the heavier gun. Army Ground Forces, on the other hand, did not object to the new tanks, but rejected the 90 mm gun as tank armament. AGF opinion was that, if the tanks were equipped with the 90 mm gun, they would be encouraged to engage in tank-versus-tank battles and abandon their proper role as a maneuvering element. In the AGF view, all antitank action was the function of the artillery and tank destroyers. This argument resulted in all production requests being rejected. Fortunately, experience overseas, where the Panthers and Tigers were being encountered in Italy, broke the deadlock. Planning for the invasion of Normandy was in full swing and General Devers cabled from London requesting the highest priority for the T26E1. In January 1944, 250 T26E1s were authorized in addition to the ten already ordered. General Barnes considered these quantities inadequate and continued to press for production of 1000 new tanks.

194

The battle against the 90 mm gun as tank armament was not yet dead and in April 1944, AGF made the ridiculous request that 6000 T25E1 and T26E1 tanks be produced with 75 mm and 76 mm guns respectively. Ordnance parried with the suggestion that such a program be given low priority, since both weapons were being supplied in the light tank M24 and the medium tank M4. Luckily, the European Theater again stepped in and, shortly before D-Day, requested that all production of 75 mm and 76 mm medium tanks be stopped. In the future, they wanted only the 90 mm gun and the 105 mm howitzer as tank armament. After D-Day and the Normandy battles with the German Panthers and Tigers, the importance of a high powered gun finally reached full realization. A less desirable result of the experience was the emphasis placed on heavier armor and the loss of interest in the T25E1. It was only natural for the troops to want as much armor protection as they could get and certainly as much as on the enemy's tanks. However, to carry the necessary weight and still retain good mobility over rough terrain required a more powerful engine than the Ford GAF. Such engines were under development

for the new heavy tanks, but were not yet in production by the spring of 1945. The T25E1 provided a much better balance between firepower, mobility, and armor with the 500 h.p. engine available.

Based on his experience, General Harmon said that the ideal tank should have the most powerful gun available, followed by the best mobility and then as much armor as possible, without interfering with the first two. Modern tank development in Europe has followed this trend, with the Leopard in Germany and the AMX 30 in France. Both vehicles emphasize firepower and mobility, with armor limited to that necessary to provide protection against automatic weapons such as the Soviet 57 mm antiaircraft guns. It is interesting to note that the T25E1 closely approached this same formula and might have been a better choice for development than its heavier brother. However, like other phases of the T20 program, it provided information for future designs and this was perhaps the greatest value of the series. The wide range of components and design concepts investigated during this period established the basis for modern tank development in this country.

Marine Pershings push north to Wonju, Korea against the Chinese Communist Forces, February 1951.

PART IV

REFERENCE DATA

THE PERSHING, PANTHER, AND TIGER I
A COMPARISON

When the Pershing was introduced into battle in early 1945, German armored power was represented by the Panther, the Tiger I, and the Tiger II. The high powered 8.8 cm KwK 43 and extremely heavy armor of the latter placed it in the category of a heavy tank, far superior in striking power and protection to the Pershing. The Pershing was not designed to compete with such a vehicle on a tank versus tank basis. The Tiger I, however, had been studied since its introduction in late 1942 and it exercised considerable influence on the final Pershing design, since one objective was to provide a medium tank with equivalent firepower and protection. The Panther, designed by the Germans after the encounter with the Soviet T34, had been developed through three models into a superb fighting vehicle intended to become the stand-

Panzerkampfwagen V, Ausfuehrung G, Panther

Medium Tank M26, General Pershing

ard of the Panzer troops. Major production plans were centered around the Panther and further development was in progress. If the war had continued, the Panther would have accounted for a much higher percentage of the German armored force. It is interesting to compare the Pershing with the Panther as well as the Tiger I, the tank it was designed to defeat. The major characteristics of all three tanks are listed in the tables and some of the more important items are covered in the following brief discussion.

The Pershing, Panther, and Tiger I were close enough in fighting power so that each could defeat the others under favorable circumstances. As described in Part I of this book, the Pershing on occasion destroyed both German vehicles and was in turn knocked out by them. The purpose of this compari-

son is to determine their relative standing in regard to the various tank design parameters. Any such comparative evaluation must first consider the basic design factors of firepower, mobility, and protection. Other qualities such as reliability, ease of transport, and crew comfort certainly have an important effect, but once committed to battle, the survival of the tank rests mainly on the three prime considerations.

In tank versus tank combat, firepower, as exemplified by armor penetration performance, is of the greatest importance. The ability of the main weapon to destroy the enemy tank at long range from any angle simplifies the tactical problem and reduces the requirements for maneuver and armor protection. Closely tied to the requirement for a powerful gun is the necessity for adequate fire control equipment to permit rapid and accurate engagement of the enemy at the maximum effective range.

Panzerkampfwagen VI, Ausfuehrung E, Tiger I

All three tanks mounted powerful weapons and their armor piercing performance overlapped somewhat, with the variety of ammunition available. The 90 mm gun M3 of the Pershing and the 8.8 cm KwK 36 of the Tiger I gave quite similar results when using the same type of APCBC ammunition with a muzzle velocity of approximately 2650 ft/sec. The Panther's 7.5 cm KwK 42, with a 400 ft/sec increase in muzzle velocity, obtained greater armor penetration than either of the larger weapons despite its lighter projectile weight. The muzzle velocity of the late production 90 mm APCBC ammunition (APC M82) was increased to 2800 ft/sec, but very little was available prior to the end of the war in Europe. The higher muzzle velocity gave the 90 superiority over the Panther's gun at ranges exceeding about 1000 yards, since the velocity did not drop off as rapidly with the heavier projectile. The new 90 mm T33 shot, heat treated to a higher hardness level, also showed superiority, particularly at the longer ranges. Introduction of the APCR type lightweight shot with the tungsten carbide core (HVAP T30E16) also increased the performance of the 90. Similar ammunition was developed for the German weapons, but production was halted due to shortages of tungsten. In spite of the great variety of results obtainable with different types of ammunition, the fact remains that 7.5 cm KwK 42 showed superior armor penetration using the standard APCBC projectiles readily available during the latter months of the war. The complete APCBC round for the Panther's gun weighed about 31 pounds compared to 43 and 34 pounds for those of the Pershing and Tiger I respectively. The lighter weight made the ammunition easier to handle in the tank turret, increasing the rate of fire. The larger guns could, of course, fire a higher capacity high explosive shell, but they were less effective as the hole punchers required for tank versus tank action. On this basis the Panther must be rated first in regard to firepower followed by the Pershing and then the Tiger I.

The power to weight ratio and the unit ground pressure provide two readily obtainable measures of mobility for an armored fighting vehicle. On the first count, the Panther was clearly superior with a ratio of 13.8 gross horsepower per short ton compared to approximately 11 horsepower per ton for the Pershing and Tiger I. On the basis of ground pressure, the Pershing and Panther were equal with a unit pressure of 12.5 psi compared to 13.9 psi for the Tiger I. Even the 13.9 psi value was obtainable only through the use of 28.5 inch wide battle tracks which extended the overall vehicle width to 147 inches. In comparison, the overall width of the Pershing and Panther were 138 and 129 inches respectively. The smaller width of the Panther was a definite advantage in negotiating narrow streets and bridges. In regard to the latter, the overall weight of the tank was important. The Pershing had the advantage of being almost four tons lighter than the Panther and over 16 tons lighter than the Tiger I. Considering all of these factors, the Panther once again must be rated first followed by the Pershing and the Tiger I.

Good practice in the design of armor requires balanced protection. This means that all parts of a vehicle should have equal protection from attack from any one direction and, if there is any variation, the greatest protection should be applied to the area most likely to be hit. Both the Pershing and the Panther violated this principle. The sloped front hull armor on both tanks was far stronger than the gun shield or turret front. Thus the weakest area from the front was the turret, which also had the highest probability of being hit. The actual thickness of the gun shield was quite close for all three tanks giving approximately the same degree of protection. The lower height of the Pershing was a decided advantage over the Panther and Tiger, reducing the chance of hits on the turret. Considering only the weakest areas from each direction, the three tanks were approximately equal in protection from the front and the Pershing and Tiger I were slightly superior from the sides and rear. One advantage of the Panther design should also be mentioned. The Panther turret was set back on the hull far enough so that a shot which ricocheted from the glacis plate would miss the gun shield. A similar ricochet on the Pershing would strike the gun shield.

The reader can make further comparisons using the data in the tables. However, based on the criteria of firepower, mobility, and protection, the Panther would have to be ranked first followed by the Pershing and lastly the Tiger I.

	Medium Tank M26 Pershing	PzKpfw V, ausf. G Panther	PzKpfw VI, ausf. E Tiger I
GENERAL DATA			
Crew	5	5	5
Length: Gun forward	340.5 inches (865cm)	348.8 inches (886cm)	333 inches (846cm)
Length: Without gun	249.1 inches (633cm)	273 inches (693.5cm)	249 inches (632 cm)
Gun Overhang:	91.4 inches (232 cm)	75.8 inches (192.5cm)	84 inches (214cm)
Width: Overall	138.3 inches (351cm)	128.8 inches (327cm)	147 inches (373cm)
Height:	109.4 inches (278cm)	118 inches (299.5cm)	114 inches (290cm)
Tread:	110 inches (279cm)	103 inches (268cm)	111 inches (282cm)
Ground Clearance:	17.2 inches (44cm)	22 inches (56cm)	17 inches (43cm)
Fire Height:	78 inches (198cm)	91 inches (230cm)	73 inches (185cm)
Turret Ring Diameter: (inside)	69 inches (175cm)	65 inches (165cm)	73 inches (185cm)
Weight, Combat Loaded:	92,355 pounds	99,873 pounds	125,400 pounds
Power to Weight Ratio: Gross	10.8 hp/ton	13.8 hp/ton	11.0 hp/ton
Ground Pressure: 0 penetration	12.5 psi	12.5 psi	13.9 psi
ARMOR			
Type:	Homogeneous, rolled & cast	Homogeneous, rolled	Homogeneous, rolled
Hull Thickness:			
Front, upper	4.0 inches (102mm) @ 46 deg.	3.1 inches (80mm) @ 55 deg.	4.0 inches (102mm) @ 24 deg.
lower	3.0 inches (76mm) @ 53 deg.	3.1 inches (80mm) @ 35 deg.	2.4 inches (62mm) @ 60-80 deg.
Sides, upper front	3.0 inches (76mm) @ 0 deg.	1.6 inches (40mm) @ 30 deg.	3.2 inches (82mm) @ 0 deg.
upper rear	2.0 inches (51mm) @ 0 deg.	1.6 inches (40mm) @ 30 deg.	3.2 inches (82mm) @ 0 deg.
lower front	3.0 inches (76mm) @ 0 deg.	1.6 inches (40mm) @ 0 deg.	2.4 inches (62mm) @ 0 deg.
lower rear	2.0 inches (51mm) @ 0 deg.	1.6 inches (40mm) @ 0 deg.	2.4 inches (62mm) @ 0 deg.
Rear	2.0 inches (51mm) @ 10 deg.	1.6 inches (40mm) @ 30 deg.	3.2 inches (82mm) @ 8 deg.
Top	0.875 ins. (22mm) @ 90 deg.	0.6 inches (15mm) @ 90 deg.	1.0 inches (26mm) @ 90 deg.
Floor, front	1.0 inches (25mm) @ 90 deg.	0.8 inches (20mm) @ 90 deg.	1.0 inches (26mm) @ 90 deg.
rear	0.5 inches (13mm) @ 90 deg.	0.5 inches (13mm) @ 90 deg.	
Turret Thickness:			
Gun Shield	4.5 inches (114mm) @ 0 deg.	4.7 inches (120mm) @ 0 deg.	4.3 inches (110mm) @ 0 deg.
Front	4.0 inches (102mm) @ 0 deg.	3.9 inches (100mm) @ 10 deg.	4.0 inches (102mm) @ 10 deg.
Sides	3.0 inches (76mm) @ 0 deg.	1.8 inches (45mm) @ 25 deg.	3.2 inches (82mm) @ 0 deg.
Rear	3.0 inches (76mm) @ 0 deg.	1.8 inches (45mm) @ 28 deg.	3.2 inches (82mm) @ 0 deg.
Top	1.0 inches (25mm) @ 90 deg.	0.6 inches (15mm) @ 84-90 deg.	1.0 inches (26mm) @ 81 deg.
ARMAMENT			
Primary:	90mm Gun M3	7.5cm KwK 42	8.8cm KwK 36
Traverse	360 deg., power & manual	360 deg., power & manual	360 deg., power & manual
Elevation	+20 to -10 degrees, manual	+20 to -4 degrees, manual	+17 to -6.5 degrees, manual
Secondary:	(1) .30 cal. MG coaxial	(1) 7.92 MG coaxial	(1) 7.92 MG coaxial
	(1) .30 cal. MG bow mount	(1) 7.92 MG bow mount	(1) 7.92 MG bow mount
	(1) .50 cal. MG AA mount	(1) 7.92 MG AA mount	(1) 7.92 MG AA mount
AMMUNITION			
Primary:	70 rounds 90mm	82 rounds 7.5 cm	92 rounds 8.8cm
Secondary:	5000 rounds .30 cal.	4200 rounds 7.92mm	5100 rounds 7.92mm
	550 rounds .50 cal.		
ENGINE			
Make and Model:	Ford GAF	Maybach HL 230 P30	Maybach HL 230 p45
Type:	Liquid cooled, V-8	Liquid cooled, V-12	Liquid cooled, V-12
Displacement:	1100 cubic inches (18.03L)	1457 cubic inches (23.88L)	1457 cubic inches (23.88L)
Horsepower: Gross	500 hp @ 2600 rpm	690 hp (700PS) @ 3000 rpm	690 hp (700 PS) @ 3000 rpm
POWER TRAIN			
Transmission:	Torqmatic	ZF AK 7-200	Maybach Olvar
Steering:	Controlled differential	MAN, multigeared	Henschel L600C
RUNNING GEAR			
Suspension:	Torsion bar	Double torsion bar	Torsion bar
Tracks: width	24 inches (61cm)	26 inches (66cm)	28.5 inches (72.5cm)
ground contact length	153.5 inches (390cm)	154.3 inches (392cm)	158 inches (401cm)
shoes/track	82	87	95
PERFORMANCE			
Maximum Speed: road	25 miles/hour	28 miles/hour	23 miles/hour
Maximum Grade:	60 per cent	60 per cent	75 per cent
Maximum Trench:	8 feet (245cm)	8 feet (245cm)	7.5 feet (229cm)
Maximum Vertical Wall:	46 inches (117cm)	35.5 inches (90cm)	31 inches (79cm)
Maximum Fording Depth:	48 inches (122cm)	75 inches (190cm)	48 (122cm) or 156 (396cm) ins.
Cruising Range: roads	100 miles, approx.	124 miles	73 miles

PzKpfw V ausf. G, Panther

Medium Tank M26 (T26E3)

PzKpfw VI ausf. E, Tiger I

The low, wide silhouette of the Pershing is obvious particularly when compared with the Panther. The gun overhang was also much greater on the Pershing than for the other two tanks.

Specifications for each of the major vehicles described in this book are outlined in the following data sheets. An effort has been made to obtain comparable values for each item despite the wide variations found in the official documents. Dimensions such as length, width, and height varied widely depending on the source. Lengths, for example, often did not indicate the position of the gun or if sandshields and a towing pintle were included. The height measured in the field varied with the load and the resulting spring compression. Where possible, the original design dimensions are used, obtained from the arsenal drawings. If these were not available, a comparable value was estimated using the measurements taken during tests at Aberdeen Proving Ground and Fort Knox.

Most of the terms used in the data sheets are self explanatory, however, some may be clarified by further discussion. For example, the tread is the width measured between track centers. The fire height is the distance measured from the ground to the centerline of the main weapon bore, when the weapon is aimed forward at zero elevation.

Engine power is listed as gross horsepower and net horsepower. The gross horsepower refers to the maximum power delivered at the output shaft of the engine including only those accessories essential to the operation of the engine. The effect of fans, generators, air cleaners, etc. is excluded from these measurements. The net horsepower is the power actually delivered at the output shaft as installed in the vehicle including all of the accessories. The same criteria apply to the measurement of the gross and net torque.

The ground contact length at zero penetration is defined as the distance between the centers of the first and last road wheels. This value is used to compute the ground contact area for determination of the ground pressure. Combat weights are used for all calculations.

The stowage of the experimental vehicles varied during their life reflecting the results of the testing program and changing user requirements. Early pilot tanks were fitted with a .30 caliber antiaircraft machine gun. This was later replaced by the .50 caliber weapon standard on all U.S. tanks.

Ballistic tests on the Pershing showed the drivers' auxiliary periscopes to be vulnerable to the blast from high explosive shells detonating against the turret. To strengthen the hull roof against this type of attack, the auxiliary periscopes were eliminated from the late production vehicles. As can be seen in photographs of the T26E5 and other late production tanks, only the hatch cover periscope are retained.

GENERAL DATA

Crew:	5 men
Length: Gun forward, over sandshields	294 inches
Length: Gun to rear, over sandshields	245 inches
Length: Over sandshields, w/o gun	227 inches
Gun Overhang: Gun forward	67 inches
Width: Over sandshields	123 inches
Height: Over mg mount	96 inches
Tread:	102 inches
Ground Clearance:	17 inches
Fire Height:	75 inches
Turret Ring Diameter: (inside)	69 inches
Weight, Combat Loaded:	65,758 pounds
Weight, Unstowed:	60,704 pounds
Power to Weight Ratio: Net	13.7 hp/ton
Gross	15.2 hp/ton
Ground Pressure: Zero penetration	13.5 psi

ARMOR

Type: Turret, cast homogeneous steel; Hull, rolled and cast homogeneous steel; Welded assembly

Hull Thickness:	Actual	Angle w/Vertical
Front, Upper	2.5 inches	47 degrees
Lower	2.5 inches	53 degrees
Sides	2.0 inches	0 degrees
Rear	1.5 inches	10 degrees
Top	0.75 inches	90 degrees
Floor, Front	1.0 inches	90 degrees
Rear	0.5 inches	90 degrees

Turret Thickness:		
Front	3.5 inches	0 degrees
Sides	2.5 inches	0 to 13 degrees
Rear	2.5 inches	0 degrees
Top	0.75 inches	90 degrees

ARMAMENT

Primary: 76 mm Gun M1A1 in Mount T79 in turret

Traverse: Hydraulic and manual	360 degrees
Traverse Rate: (max)	15 seconds/360 degrees
Elevation: Manual	+25 to -10 degrees
Firing Rate: (max)	20 rounds/minute
Loading System:	Manual
Stabilizer System:	Elevation only

Secondary:
(1) .30 caliber MG M1919A4 flexible AA mount on turret*
(1) .30 caliber MG M1919A4 coaxial w/76mm gun in turret
(1) .30 caliber MG M1919A4 in bow mount
Provision for (1) .45 caliber SMG M1928A1

AMMUNITION

70 rounds 76mm	12 hand grenades
540 rounds .45 caliber	
7000 rounds .30 caliber	

FIRE CONTROL AND VISION EQUIPMENT

Primary Weapon:	Direct	Indirect
	Telescope T92	Azimuth Indicator M19
	Periscope M4	Gunner's Quadrant M1
	with telescope	Elevation Quadrant M9

Vision Devices:	Direct	Indirect
Driver	Hatch	Periscope M6 (2)
Asst. Driver	Hatch	Periscope M6 (2)
Commander	Hatch	Periscope M6 (1)
Gunner	None	Periscope M4 (1)
Loader	Hatch	Periscope M6 (1)

Total Periscopes: M4 (1), M6 (6)

*Early armament and stowage, .30 caliber AA MG later replaced by a .50 caliber weapon

ENGINE

Make and Model: Ford GAN	
Type: 8 cylinder, 4 cycle, 60 degree vee	
Cooling System: Liquid Ignition: Magneto	
Displacement:	1100 cubic inches
Bore and Stroke:	5.4 x 6 inches
Compression Ratio:	7.5:1
Net Horsepower (max):	450 hp at 2600 rpm
Gross Horsepower (max):	500 hp at 2600 rpm
Net Torque (max):	950 ft-lb at 2200 rpm
Gross Torque (max):	1040 ft-lb at 2200 rpm
Weight:	1414 lb, dry
Fuel: 80 octane gasoline	196 gallons
Engine Oil:	32 quarts

POWER TRAIN

Transmission: Torqmatic, 3 speeds forward, 1 reverse
    Torque Converter Ratio: Varies from 1:1 to 4.4:1

Gear Ratios:	1st	1.63:1	3rd	0.39:1
	2nd	0.74:1	reverse	1.17:1

Steering: Controlled differential
    Bevel Gear Ratio: 2.434:1   Steering Ratio: 1.515:1
Brakes: Mechanical, external contracting
Final Drive: Spur gear   Gear Ratio: 2.84:1
Drive Sprocket: At rear of vehicle with 13 teeth
    Pitch Diameter: 25.038 inches

RUNNING GEAR

Suspension: Horizontal volute spring
    12 wheels in 6 bogies (3 bogies/track)
    Tire Size: 20 x 9 inches
    6 track return rollers (3/track)
    Adjustable idler at front of each track
    Idler Size: 22 x 9 inches
    Shock absorbers fitted on front and rear bogies
Tracks: Outside guide, T48 and T51
    Type: (T48) Double pin, 16.56 inch width, chevron
        (T51) Double pin, 16.56 inch width, smooth
    Pitch: 6 inches
    Shoes per Vehicle: 158 (79/track)
    Ground Contact Length: 147 inches

ELECTRICAL SYSTEM

Nominal Voltage: 24 volts DC
Main Generator: (2) 24 volts, 50 amperes, driven by power take-off from main engine
Auxiliary Generator: (1) 24 volts, 125 amperes, driven by the auxiliary engine
Battery: (2) 12 volts in series

COMMUNICATIONS

Radio: SCR 508 or 528 in rear of turret; SCR 506 (command tanks only) on shelf in front of loader
Interphone: (part of radio) 5 stations
Flag Set M238, Panel Set AP50A, Spotlight
Flares: 3 each, M17, M18, M19, and M21 (command tanks only)
Ground Signals Projector M4 (command tanks only)

FIRE AND GAS PROTECTION

(2) 10 pound carbon dioxide, fixed
(2) 4 pound carbon dioxide, portable
(2) 1½ quart decontaminating apparatus

PERFORMANCE

Maximum Speed: Sustained, level road	30 miles/hour
Short periods, level	35 miles/hour
Maximum Tractive Effort: TE at stall	40,000 pounds
Per Cent of Vehicle Weight: TE/W	60.8 per cent
Maximum Grade:	58 per cent
Maximum Trench:	7.5 feet
Maximum Vertical Wall:	24 inches
Maximum Fording Depth:	48 inches
Minimum Turning Circle: (diameter)	65 feet
Cruising Range: Roads approx.	150 miles

MEDIUM TANK T20E3

GENERAL DATA

Crew:	5 men
Length: Gun forward, over sandshields	294 inches
Length: Gun to rear, over sandshields	245 inches
Length: Over sandshields, w/o gun	227 inches
Gun Overhang: Gun forward	67 inches
Width: Over sandshields	127 inches
Height: Over mg mount	97 inches
Tread:	105 inches
Ground Clearance:	16 inches
Fire Height:	76 inches
Turret Ring Diameter: (inside)	69 inches
Weight, Combat Loaded:	67,500 pounds
Weight, Unstowed:	62,500 pounds
Power to Weight Ratio: Net	13.3 hp/ton
Gross	14.8 hp/ton
Ground Pressure: Zero penetration	12.8 psi

ARMOR

Type: Turret, cast homogeneous steel; Hull, rolled and cast homogeneous steel; Welded assembly

Hull Thickness:		Actual		Angle w/Vertical
Front, Upper		2.5	inches	47 degrees
Lower		2.5	inches	53 degrees
Sides		2.0	inches	0 degrees
Rear		1.5	inches	10 degrees
Top		0.75	inches	90 degrees
Floor, Front		1.0	inches	90 degrees
Rear		0.5	inches	90 degrees
Turret Thickness:				
Front		3.5	inches	0 degrees
Sides		2.5	inches	0 to 13 degrees
Rear		2.5	inches	0 degrees
Top		0.75	inches	90 degrees

ARMAMENT

Primary: 76mm Gun M1A1 in Mount T79 in turret

Traverse: Hydraulic and manual	360 degrees
Traverse Rate: (max)	15 seconds/360 degrees
Elevation: Manual	+25 to -10 degrees
Firing Rate: (max)	20 rounds/minute
Loading System:	Manual
Stabilizer System:	Elevation only

Secondary:
   (1) .50 caliber MG HB M2 flexible AA mount on turret*
   (1) .30 caliber MG M1919A4 coaxial w/76mm gun in turret
   (1) .30 caliber MG M1919A4 in bow mount
   Provision for (1) .45 caliber SMG M1928A1

AMMUNITION

68 rounds 76mm	12 hand grenades
300 rounds .50 caliber	
600 rounds .45 caliber	
5000 rounds .30 caliber	

FIRE CONTROL AND VISION EQUIPMENT

Primary Weapon:	Direct	Indirect
	Telescope T92	Azimuth Indicator M19
	Periscope M4	Gunner's Quadrant M1
	with telescope	Elevation Quadrant M9
Vision Devices:	Direct	Indirect
Driver	Hatch	Periscope M6 (2)
Asst. Driver	Hatch	Periscope M6 (2)
Commander	Hatch	Periscope M6 (1)
Gunner	None	Periscope M4 (1)
Loader	Hatch	Periscope M6 (1)

Total Periscopes: M4 (1), M6 (6)

*Late armament and stowage, a .30 caliber AA MG was originally specified

ENGINE

Make and Model: Ford GAN	
Type: 8 cylinder, 4 cycle, 60 degree vee	
Cooling System: Liquid Ignition: Magneto	
Displacement:	1100 cubic inches
Bore and Stroke:	5.4 x 6 inches
Compression Ratio:	7.5:1
Net Horsepower (max):	450 hp at 2600 rpm
Gross Horsepower (max):	500 hp at 2600 rpm
Net Torque (max):	950 ft-lb at 2200 rpm
Gross Torque (max):	1040 ft-lb at 2200 rpm
Weight:	1414 lb, dry
Fuel: 80 octane gasoline	196 gallons
Engine Oil:	32 quarts

POWER TRAIN

Transmission: Torqmatic, 3 speeds forward, 1 reverse
   Torque Converter Ratio: Varies from 1:1 to 4.4:1

Gear Ratios:	1st	1.63:1	3rd	0.39:1
	2nd	0.74:1	reverse	1.17:1

Steering: Controlled differential
   Bevel Gear Ratio: 2.434:1   Steering Ratio: 1.515:1
Brakes: Mechanical, external contracting
Final Drive: Spur gear   Gear Ratio: 2.84:1
Drive Sprocket: At rear of vehicle with 27 teeth
   Pitch Diameter: 25.054 inches

RUNNING GEAR

Suspension: Torsion bar
   12 individually sprung dual road wheels (6/track)
   Tire Size: 26 x 4.5 inches
   6 dual track return rollers (3/track)
   Final version equipped with 10 track return rollers (5/track)
   Dual compensating idler at front of each track
   Idler Size: 24.5 x 4.5 inches
   Shock absorbers fitted on first 2 and last 2 road wheels on each side
Tracks: Center guide, cast steel
   Type: Single pin, 18 inch width
   Pitch: 6 inches
   Shoes per Vehicle: 166 (83/track)
   Ground Contact Length: 145.2 inches, left side
                         148.7 inches, right side

ELECTRICAL SYSTEM

Nominal Voltage: 24 volts DC
Main Generator: (2) 24 volts, 50 amperes, driven by power take-off from main engine
Auxiliary Generator: (1) 24 volts, 125 amperes, driven by the auxiliary engine
Battery: (2) 12 volts in series

COMMUNICATIONS

Radio: SCR 508 or 528 in rear of turret; SCR 506 (command tanks only) on shelf in front of loader
Interphone: (part of radio) 5 stations
Flag Set M238, Panel Set AP50A, Spotlight
Flares: 3 each, M17, M18, M19, and M21 (command tanks only)
Ground Signals Projector M4 (command tanks only)

FIRE AND GAS PROTECTION

   (2) 10 pound carbon dioxide, fixed
   (2) 4 pound carbon dioxide, portable
   (2) 1½ quart decontaminating apparatus

PERFORMANCE

Maximum Speed: Sustained, level road		30 miles/hour
Short periods, level		35 miles/hour
Maximum Tractive Effort: TE at stall		40,000 pounds
Per Cent of Vehicle Weight: TE/W		59.3 per cent
Maximum Grade:		50 per cent
Maximum Trench:		7.5 feet
Maximum Vertical Wall:		24 inches
Maximum Fording Depth:		48 inches
Minimum Turning Circle: (diameter)		65 feet
Cruising Range: Roads	approx.	150 miles

MEDIUM TANK T22

GENERAL DATA

Crew:	5 men
Length: Gun forward, over sandshields	293 inches
Length: Gun to rear, over sandshields	253 inches
Length: Over sandshields, w/o gun	240 inches
Gun Overhang: Gun forward	53 inches
Width: Over sandshields	123 inches
Height: Over mg mount	96 inches
Tread:	102 inches
Ground Clearance:	19 inches
Fire Height:	77 inches
Turret Ring Diameter: (inside)	69 inches
Weight, Combat Loaded:	69,300 pounds
Weight, Unstowed:	64,246 pounds
Power to Weight Ratio: Net	13.0 hp/ton
Gross	14.4 hp/ton
Ground Pressure: Zero penetration	14.2 psi

ARMOR

Type: Turret, cast homogeneous steel; Hull, rolled and cast homogeneous steel; Welded assembly

Hull Thickness:

		Actual	Angle w/Vertical
Front, Upper		2.5 inches	47 degrees
Lower		2.5 inches	53 degrees
Sides		2.0 inches	0 degrees
Rear		1.5 inches	10 degrees
Top		0.75 inches	90 degrees
Floor, Front		1.0 inches	90 degrees
Rear		0.5 inches	90 degrees

Turret Thickness:

	Actual	Angle w/Vertical
Front	3.5 inches	0 degrees
Sides	2.5 inches	0 to 13 degrees
Rear	2.5 inches	0 degrees
Top	0.75 inches	90 degrees

ARMAMENT

Primary: 76mm Gun M1A1 in Mount T79 in turret

Traverse: Hydraulic and manual	360 degrees
Traverse Rate: (max)	15 seconds/360 degrees
Elevation: Manual	+25 to -10 degrees
Firing Rate: (max)	20 rounds/minute
Loading System:	Manual
Stabilizer System:	Elevation only

Secondary:
- (1) .50 caliber MG HB M2 flexible AA mount on turret*
- (1) .30 caliber MG M1919A4 coaxial w/76mm gun in turret
- (1) .30 caliber MG M1919A4 in bow mount
- Provision for (1) .45 caliber SMG M1928A1

AMMUNITION

68 rounds 76mm	12 hand grenades
300 rounds .50 caliber	
600 rounds .45 caliber	
5000 rounds .30 caliber	

FIRE CONTROL AND VISION EQUIPMENT

Primary Weapon:	Direct	Indirect
	Telescope T92	Azimuth Indicator M19
	Periscope M4	Gunner's Quadrant M1
	with telescope	Elevation Quadrant M9

Vision Devices:	Direct	Indirect
Driver	Hatch	Periscope M6 (2)
Asst. Driver	Hatch	Periscope M6 (2)
Commander	Hatch	Periscope M6 (1)
Gunner	None	Periscope M4 (1)
Loader	Hatch	Periscope M6 (1)

Total Periscopes: M4 (1), M6 (6)

*Late armament and stowage, a .30 caliber AA MG was originally specified

ENGINE

Make and Model: Ford GAN	
Type: 8 cylinder, 4 cycle, 60 degree vee	
Cooling System: Liquid Ignition: Magneto	
Displacement:	1100 cubic inches
Bore and Stroke:	5.4 x 6 inches
Compression Ratio:	7.5:1
Net Horsepower (max):	450 hp at 2600 rpm
Gross Horsepower (max):	500 hp at 2600 rpm
Net Torque (max):	950 ft-lb at 2200 rpm
Gross Torque (max):	1040 ft-lb at 2200 rpm
Weight:	1414 lb, dry
Fuel: 80 octane gasoline	200 gallons
Engine Oil:	32 quarts

POWER TRAIN

Clutch: Dry disc, 2 plate

Transmission: Synchromesh, 5 speeds forward, 1 reverse

Gear Ratios:	1st	7.554:1	4th	1.108:1
	2nd	3.105:1	5th	0.733:1
	3rd	1.768:1	reverse	5.646:1

Steering: Controlled differential
Bevel Gear Ratio: 3.53:1   Steering Ratio: 1.515:1
Brakes: Mechanical, external contracting
Final Drive: Spur gear   Gear Ratio: 2.84:1
Drive Sprocket: At rear of vehicle with 13 teeth
Pitch Diameter: 25.038 inches

RUNNING GEAR

Suspension: Horizontal volute spring
12 wheels in 6 bogies (3 bogies/track)
Tire Size: 20 x 9 inches
6 track return rollers (3/track)
Adjustable idler at front of each track
Idler Size: 22 x 9 inches
Shock absorbers fitted on front and rear bogies
Tracks: Outside guide, T48 and T51
Type: (T48) Double pin, 16.56 inch width, chevron
(T51) Double pin, 16.56 inch width, smooth
Pitch: 6 inches
Shoes per Vehicle: 158 (79/track)
Ground Contact Length: 147 inches

ELECTRICAL SYSTEM

Nominal Voltage: 24 volts DC
Main Generator: (1) 24 volts, 50 amperes, driven by power take-off from main engine
Auxiliary Generator: (1) 24 volts, 125 amperes, driven by the auxiliary engine
Battery: (2) 12 volts in series

COMMUNICATIONS

Radio: SCR 508 or 528 in rear of turret; SCR 506 (command tanks only) on shelf in front of loader
Interphone: (part of radio) 5 stations
Flag Set M238, Panel Set AP50A, Spotlight
Flares: 3 each, M17, M18, M19, and M21 (command tanks only)
Ground Signals Projector M4 (command tanks only)

FIRE AND GAS PROTECTION

(2) 10 pound carbon dioxide, fixed
(2) 4 pound carbon dioxide, portable
(2) 1½ quart decontaminating apparatus

PERFORMANCE

Maximum Speed: Sustained, level road	25 miles/hour
Maximum Tractive Effort: TE at stall	57,600 pounds
Per Cent of Vehicle Weight: TE/W	83.1 per cent
Maximum Grade:	60 per cent
Maximum Trench:	7.5 feet
Maximum Vertical Wall:	24 inches
Maximum Fording Depth:	48 inches
Minimum Turning Circle: (diameter)	65 feet
Cruising Range: Roads approx.	150 miles

MEDIUM TANK T22E1

GENERAL DATA

Crew:		4 men
Length: Gun forward, over sandshields		254 inches
Length: Gun to rear, over sandshields		240 inches
Length: Over sandshields, w/o gun		240 inches
Gun Overhang: Gun forward		14 inches
Width: Over sandshields		123 inches
Height: Over mg mount		95 inches
Tread:		102 inches
Ground Clearance:		19 inches
Fire Height:		77 inches
Turret Ring Diameter: (inside)		69 inches
Weight, Combat Loaded:	estimated	68,000 pounds
Weight, Unstowed:	estimated	63,000 pounds
Power to Weight Ratio: Net		13.2 hp/ton
Gross		14.7 hp/ton
Ground Pressure: Zero penetration		14.0 psi

ARMOR

Type: Turret, cast homogeneous steel; Hull, rolled and cast homogeneous steel; Welded assembly

Hull Thickness:	Actual	Angle w/Vertical
Front, Upper	2.5 inches	47 degrees
Lower	2.5 inches	53 degrees
Sides	2.0 inches	0 degrees
Rear	1.5 inches	10 degrees
Top	0.75 inches	90 degrees
Floor, Front	1.0 inches	90 degrees
Rear	0.5 inches	90 degrees
Turret Thickness:		
Front	3.5 inches	0 degrees
Sides	2.5 inches	0 degrees
Rear	2.5 inches	0 degrees
Top	1.0 inches	90 degrees

ARMAMENT

Primary: 75mm Gun M3 in Mount M34 in turret

Traverse: Hydraulic and manual	360 degrees
Traverse Rate: (max)	15 seconds/360 degrees
Elevation: Manual	+15 to -10 degrees
Firing Rate: (max)	20 rounds/minute
Loading System:	Automatic, hydraulic
Stabilizer System:	Elevation only

Secondary:
(1) .50 caliber MG HB M2 flexible AA mount on turret*
(1) .30 caliber MG M1919A4 coaxial w/75mm gun in turret
(1) .30 caliber MG M1919A4 in bow mount
Provision for (1) .45 caliber SMG M1928A1

AMMUNITION

64 rounds 75mm	12 hand grenades
300 rounds .50 caliber	
600 rounds .45 caliber	
5000 rounds .30 caliber	

FIRE CONTROL AND VISION EQUIPMENT

Primary Weapon:	Direct	Indirect
	Periscope M4	Azimuth Indicator M19
	with Telescope	Gunner's Quadrant M1
	M38	Elevation Quadrant M9
Vision Devices:	Direct	Indirect
Driver	Hatch	Periscope M6 (2)
Asst. Driver	Hatch	Periscope M6 (2)
Commander	Hatch and	Periscope M6 (1)
	pistol port	
Gunner	Hatch	Periscope M4 (1)

Total Periscopes: M4 (1), M6 (5)
Total Pistol Ports: Hull (0), Turret (1)

*Late armament and stowage, a .30 caliber AA MG was originally specified

ENGINE

Make and Model: Ford GAN	
Type: 8 cylinder, 4 cycle, 60 degree vee	
Cooling System: Liquid Ignition: Magneto	
Displacement:	1100 cubic inches
Bore and Stroke:	5.4 x 6 inches
Compression Ratio:	7.5:1
Net Horsepower (max):	450 hp at 2600 rpm
Gross Horsepower (max):	500 hp at 2600 rpm
Net Torque (max):	950 ft-lb at 2200 rpm
Gross Torque (max):	1040 ft-lb at 2200 rpm
Weight:	1414 lb, dry
Fuel: 80 octane gasoline	200 gallons
Engine Oil:	32 quarts

POWER TRAIN

Clutch: Dry disc, 2 plate
Transmission: Synchromesh, 5 speeds forward, 1 reverse

Gear Ratios:	1st	7.554:1	4th	1.108:1
	2nd	3.105:1	5th	0.733:1
	3rd	1.768:1	reverse	5.646:1

Steering: Controlled differential
Bevel Gear Ratio: 3.53:1   Steering Ratio: 1.515:1
Brakes: Mechanical, external contracting
Final Drive: Spur gear   Gear Ratio: 2.84:1
Drive Sprocket: At rear of vehicle with 13 teeth
Pitch Diameter: 25.038 inches

RUNNING GEAR

Suspension: Horizontal volute spring
12 wheels in 6 bogies (3 bogies/track)
Tire Size: 20 x 9 inches
6 track return rollers (3/track)
Adjustable idler at front of each track
Idler Size: 22 x 9 inches
Shock absorbers fitted on front and rear bogies
Tracks: Outside guide, T48 and T51
Type: (T48) Double pin, 16.56 inch width, chevron
(T51) Double pin, 16.56 inch width, smooth
Pitch: 6 inches
Shoes per Vehicle: 158 (79/track)
Ground Contact Length: 147 inches

ELECTRICAL SYSTEM

Nominal Voltage: 24 volts DC
Main Generator: (1) 24 volts, 50 amperes, driven by power take-off from main engine
Auxiliary Generator: (1) 24 volts, 125 amperes, driven by the auxiliary engine
Battery: (2) 12 volts in series

COMMUNICATIONS

Radio: SCR 508 or 528 in rear of turret; SCR 506 (command tanks only) on shelf in front of loader
Interphone: (part of radio) 4 stations
Flag Set M238, Panel Set AP50A, Spotlight
Flares: 3 each, M17, M18, M19, and M21 (command tanks only)
Ground Signals Projector M4 (command tanks only)

FIRE AND GAS PROTECTION

(2) 10 pound carbon dioxide, fixed
(1) 4 pound carbon dioxide, portable
(2) 1½ quart decontaminating apparatus

PERFORMANCE

Maximum Speed: Sustained, level road	25 miles/hour
Maximum Tractive Effort: TE at stall	57,600 pounds
Per Cent of Vehicle Weight: TE/W	84.7 per cent
Maximum Grade:	60 per cent
Maximum Trench:	7.5 feet
Maximum Vertical Wall:	24 inches
Maximum Fording Depth:	48 inches
Minimum Turning Circle: (diameter)	62 feet
Cruising Range: Roads approx.	150 miles

MEDIUM TANK T23, Pilot #2

GENERAL DATA

Crew:	5 men
Length: Gun forward, over sandshields	298 inches
Length: Gun to rear, over sandshields	254 inches
Length: Over sandshields, w/o gun	237 inches
Gun Overhang: Gun forward	61 inches
Width: Over sandshields	123 inches
Height: Over mg mount	96 inches
Tread:	102 inches
Ground Clearance:	19 inches
Fire Height:	77 inches
Turret Ring Diameter: (inside)	69 inches
Weight, Combat Loaded:	72,500 pounds
Weight, Unstowed:	66,500 pounds
Power to Weight Ratio: Net	12.4 hp/ton
Gross	13.8 hp/ton
Ground Pressure: Zero penetration	14.9 psi

ARMOR

Type: Turret, cast homogeneous steel; Hull, rolled and cast homogeneous steel; Welded assembly

Hull Thickness:	Actual	Angle w/Vertical
Front, Upper	2.5 inches	47 degrees
Lower	2.0 inches	53 degrees
Sides	2.0 inches	0 degrees
Rear	1.5 inches	0 to 30 degrees
Top	0.75 inches	90 degrees
Floor, Front	1.0 inches	90 degrees
Rear	0.5 inches	90 degrees
Turret Thickness:		
Front	3.5 inches	0 degrees
Sides	2.5 inches	0 to 13 degrees
Rear	2.5 inches	0 degrees
Top	0.75 inches	90 degrees

ARMAMENT

Primary: 76mm Gun M1A1 in Mount T79 in turret

Traverse: Hydraulic and manual	360 degrees
Traverse Rate: (max)	15 seconds/360 degrees
Elevation: Manual	+25 to -10 degrees
Firing Rate: (max)	20 rounds/minute
Loading System:	Manual
Stabilizer System:	Elevation only

Secondary:
   (1) .30 caliber MG M1919A4 flexible AA mount on turret*
   (1) .30 caliber MG M1919A4 coaxial w/76mm gun in turret
   (1) .30 caliber MG M1919A4 in bow mount
   Provision for (1) .45 caliber SMG M1928A1

AMMUNITION

75 rounds 76mm	12 hand grenades
450 rounds .45 caliber	
7000 rounds .30 caliber	

FIRE CONTROL AND VISION EQUIPMENT

Primary Weapon:	Direct	Indirect
	Telescope T92	Azimuth Indicator M19
	Periscope M4	Gunner's Quadrant M1
	with telescope	Elevation Quadrant M9

Vision Devices:	Direct	Indirect
Driver	Hatch	Periscope M6 (2)
Asst. Driver	Hatch	Periscope M6 (2)
Commander	Hatch	Periscope M6 (1)
Gunner	None	Periscope M4 (1)
Loader	Hatch	Periscope M6 (1)

Total Periscopes: M4 (1), M6 (6)

*Early armament and stowage, .30 caliber AA MG later replaced by a .50 caliber weapon

ENGINE

Make and Model: Ford GAN	
Type: 8 cylinder, 4 cycle, 60 degree vee	
Cooling System: Liquid Ignition: Magneto	
Displacement:	1100 cubic inches
Bore and Stroke:	5.4 x 6 inches
Compression Ratio:	7.5:1
Net Horsepower (max):	450 hp at 2600 rpm
Gross Horsepower (max):	500 hp at 2600 rpm
Net Torque (max):	950 ft-lb at 2200 rpm
Gross Torque (max):	1040 ft-lb at 2200 rpm
Weight:	1414 lb, dry
Fuel: 80 octane gasoline	179 gallons
Engine Oil:	32 quarts

POWER TRAIN

Transmission: Electric drive with speed infinitely variable both forward and reverse
Steering: Electric
Brakes: Electric
Final Drive: Spur gear   Gear Ratio: 5.31:1
Drive Sprocket: At rear of vehicle with 13 teeth
   Pitch Diameter: 25.038 inches

RUNNING GEAR

Suspension: Vertical volute spring
   12 wheels in 6 bogies (3 bogies/track)
   Tire Size: 20 x 9 inches
   6 track return rollers (1 at rear of each bogie)
   Adjustable idler at front of each track
   Idler Size: 22 x 9 inches
Tracks: Outside guide, T48 and T51
   Type: (T48) Double pin, 16.56 inch width, chevron
       (T51) Double pin, 16.56 inch width, smooth
   Pitch: 6 inches
   Shoes per Vehicle: 158 (79/track)
   Ground Contact Length: 147 inches

ELECTRICAL SYSTEM

Nominal Voltage: 24 volts DC
Battery Charging Generator: 24 volts, 300 amperes, driven by power take-off from main engine
Auxiliary Generator: 24 volts, 50 amperes, driven by the auxiliary engine
Battery: (2) 12 volts in series

COMMUNICATIONS

Radio: SCR 508 or 528 in rear of turret; SCR 506 (command tanks only) on shelf in front of loader
Interphone: (part of radio) 5 stations
Flag Set M238, Panel Set AP50A, Spotlight
Flares: 3 each, M17, M18, M19, and M21 (command tanks only)
Ground Signals Projector M4 (command tanks only)

FIRE AND GAS PROTECTION

   (2) 10 pound carbon dioxide, fixed
   (2) 4 pound carbon dioxide, portable
   (2) 1½ quart decontaminating apparatus

PERFORMANCE

Maximum Speed: Sustained, level road	35 miles/hour
Maximum Tractive Effort: TE at stall	56,000 pounds
Per Cent of Vehicle Weight: TE/W	77.2 per cent
Maximum Grade:	60 per cent
Maximum Trench:	7.5 feet
Maximum Vertical Wall:	24 inches
Maximum Fording Depth:	48 inches
Minimum Turning Circle: (diameter)	pivot
Cruising Range: Roads approx.	150 miles

MEDIUM TANK T23

GENERAL DATA

Crew:	5	men
Length: Gun forward, over sandshields	297.5	inches
Length: Gun to rear, over sandshields	253.8	inches
Length: Over sandshields, w/o gun	236.9	inches
Gun Overhang: Gun forward	60.6	inches
Width: Over sandshields	122.5	inches
Height: To top of cupola	98.8	inches
Tread:	102	inches
Ground Clearance:	19.1	inches
Fire Height:	77	inches
Turret Ring Diameter: (inside)	69	inches
Weight, Combat Loaded:	75,311	pounds
Weight, Unstowed:	68,200	pounds
Power to Weight Ratio: Net	11.9	hp/ton
Gross	13.3	hp/ton
Ground Pressure: Zero penetration	15.5	psi

ARMOR

Type: Turret, cast homogeneous steel; Hull, rolled and cast homogeneous steel; Welded assembly

Hull Thickness:	Actual	Angle w/Vertical
Front, Upper	3.0 inches	47 degrees
Lower	2.5 inches	56 degrees
Sides, Front	2.0 inches	0 degrees
Rear	1.5 inches	0 degrees
Rear	1.5 inches	0 to 30 degrees
Top	0.75 inches	90 degrees
Floor, Front	1.0 inches	90 degrees
Rear	0.5 inches	90 degrees

Turret Thickness:		
Gun Shield	3.5 inches	0 degrees
Front	3.0 inches	0 degrees
Sides	2.5 inches	0 to 13 degrees
Rear	2.5 inches	0 degrees
Top	1.0 inches	90 degrees

ARMAMENT

Primary: 76mm Gun M1A1 in Mount M62 (T80) in turret

Traverse: Hydraulic and manual	360 degrees
Traverse Rate: (max)	15 seconds/360 degrees
Elevation: Manual	+25 to -10 degrees
Firing Rate: (max)	20 rounds/minute
Loading System:	Manual
Stabilizer System:	Elevation only

Secondary:
(1) .50 caliber MG HB M2 flexible AA mount on turret
(1) .30 caliber MG M1919A4 coaxial w/76mm gun in turret
(1) .30 caliber MG M1919A4 in bow mount
(1) 2 inch Mortar M3 (smoke) fixed in turret
Provision for (4) .45 caliber SMG M3
Provision for (1) .30 caliber Carbine M1

AMMUNITION

66 rounds 76mm	12 rounds 2 in. (smoke)
300 rounds .50 caliber	12 hand grenades
600 rounds .45 caliber	
5000 rounds .30 caliber	

FIRE CONTROL AND VISION EQUIPMENT

Primary Weapon:	Direct	Indirect
	Telescope M71D	Azimuth Indicator M19
	Periscope M4A1	Gunner's Quadrant M1
	with Telescope M47	Elevation Quadrant M9

Vision Devices:	Direct	Indirect
Driver	Hatch	Periscope M6 (2)
Asst. Driver	Hatch	Periscope M6 (2)
Commander	Vision cupola and hatch	Periscope M6 (1)
Gunner	None	Periscope M4A1 (1)
Loader	Hatch and pistol port	Periscope M6 (1)

Total Periscopes: M4A1 (1), M6 (6)
Total Pistol Ports: Hull (0), Turret (1)
Vision Cupolas: (1) w/6 vision blocks on turret top

ENGINE

Make and Model: Ford GAN	
Type: 8 cylinder, 4 cycle, 60 degree vee	
Cooling System: Liquid Ignition: Magneto	
Displacement:	1100 cubic inches
Bore and Stroke:	5.4 x 6 inches
Compression Ratio:	7.5:1
Net Horsepower (max):	450 hp at 2600 rpm
Gross Horsepower (max):	500 hp at 2600 rpm
Net Torque (max):	950 ft-lb at 2200 rpm
Gross Torque (max):	1040 ft-lb at 2200 rpm
Weight:	1414 lb, dry
Fuel: 80 octane gasoline	179 gallons
Engine Oil:	32 quarts

POWER TRAIN

Transmission: Electric drive with speed infinitely variable both forward and reverse
Steering: Electric
Brakes: Electric and mechanical Bendix
Final Drive: Spur gear   Gear Ratio: 5.31:1
Drive Sprocket: At rear of vehicle with 13 teeth
    Pitch Diameter: 25.038 inches

RUNNING GEAR

Suspension: Vertical volute spring
    12 wheels in 6 bogies (3 bogies/track)
    Tire Size: 20 x 9 inches
    6 track return rollers (1 at rear of each bogie)
    Adjustable idler at front of each track
    Idler Size: 22 x 9 inches
Tracks: Outside guide, T48 and T51
    Type: (T48) Double pin, 16.56 inch width, chevron
          (T51) Double pin, 16.56 inch width, smooth
    Pitch: 6 inches
    Shoes per Vehicle: 158 (79/track)
    Ground Contact Length: 147 inches

ELECTRICAL SYSTEM

Nominal Voltage: 24 volts DC
Battery Charging Generator: 24 volts, 200 amperes, driven by power take-off from main engine
Auxiliary Generator: 24 volts, 50 amperes, driven by the auxiliary engine
Battery: (2) 12 volts in series

COMMUNICATIONS

Radio: SCR 508 or 528 in rear of turret; SCR 506 (command tanks only) on shelf in front of loader
Interphone: (part of radio) 5 stations
Flag Set M238, Panel Set AP50A, Spotlight
Flares: 3 each, M17, M18, M19, and M21 (command tanks only)
Ground Signals Projector M4 (command tanks only)

FIRE AND GAS PROTECTION

(2) 10 pound carbon dioxide, fixed
(2) 4 pound carbon dioxide, portable
(2) 1½ quart decontaminating apparatus

PERFORMANCE

Maximum Speed: Sustained, level road	35 miles/hour
Maximum Tractive Effort: TE at stall	56,000 pounds
Per Cent of Vehicle Weight: TE/W	74.4 per cent
Maximum Grade:	60 per cent
Maximum Trench:	7.5 feet
Maximum Vertical Wall:	24 inches
Maximum Fording Depth:	48 inches
Minimum Turning Circle: (diameter)	pivot
Cruising Range: Roads approx.	100 miles

MEDIUM TANK T23E3

GENERAL DATA

Crew:	5	men
Length: Gun forward, over sandshields	306	inches
Length: Gun to rear, over sandshields	253.8	inches
Length: Over sandshields, w/o gun	244.4	inches
Gun Overhang: Gun forward	62	inches
Width: Over sandshields	128	inches
Height: To top of cupola	100.6	inches
Tread:	105	inches
Ground Clearance:	17.6	inches
Fire Height:	78	inches
Turret Ring Diameter: (inside)	69	inches
Weight, Combat Loaded:	79,390	pounds
Weight, Unstowed:	72,000	pounds
Power to Weight Ratio: Net	11.3	hp/ton
Gross	12.6	hp/ton
Ground Pressure: Zero penetration	13.9	psi

ARMOR
Type: Turret, cast homogeneous steel; Hull, rolled and cast homogeneous steel; Welded assembly

Hull Thickness:	Actual	Angle w/Vertical
Front, Upper	3.0 inches	47 degrees
Lower	2.5 inches	56 degrees
Sides, Front	2.0 inches	0 degrees
Rear	1.5 inches	0 degrees
Rear	1.5 inches	0 to 30 degrees
Top	0.75 inches	90 degrees
Floor, Front	1.0 inches	90 degrees
Rear	0.5 inches	90 degrees
Turret Thickness:		
Gun Shield	3.5 inches	0 degrees
Front	3.0 inches	0 degrees
Sides	2.5 inches	0 to 13 degrees
Rear	2.5 inches	0 degrees
Top	1.0 inches	90 degrees

ARMAMENT
Primary: 76mm Gun M1A1 in Mount M62 (T80) in turret

Traverse: Hydraulic and manual	360 degrees
Traverse Rate: (max)	15 seconds/360 degrees
Elevation: Manual	+25 to -10 degrees
Firing Rate: (max)	20 rounds/minute
Loading System:	Manual
Stabilizer System:	Elevation only

Secondary:
- (1) .50 caliber MG HB M2 flexible AA mount on turret
- (1) .30 caliber MG M1919A4 coaxial w/76mm gun in turret
- (1) .30 caliber MG M1919A4 in bow mount
- (1) 2 inch Mortar M3 (smoke) fixed in turret
- Provision for (4) .45 caliber SMG M3
- Provision for (1) .30 caliber Carbine M1

AMMUNITION

84 rounds 76mm	12 rounds 2 in. (smoke)
500 rounds .50 caliber	12 hand grenades
600 rounds .45 caliber	
4000 rounds .30 caliber	

FIRE CONTROL AND VISION EQUIPMENT

Primary Weapon:	Direct	Indirect
	Telescope M71D	Azimuth Indicator M19
	Periscope M4A1	Gunner's Quadrant M1
	with Telescope M47	Elevation Quadrant M9
Vision Devices:	Direct	Indirect
Driver	Hatch	Periscope M6 (2)
Asst. Driver	Hatch	Periscope M6 (2)
Commander	Vision cupola and hatch	Periscope M6 (1)
Gunner	None	Periscope M4A1 (1)
Loader	Hatch and pistol port	Periscope M6 (1)

Total Periscopes: M4A1 (1), M6 (6)
Total Pistol Ports: Hull (0), Turret (1)
Vision Cupolas: (1) w/6 vision blocks on turret top

ENGINE

Make and Model: Ford GAN	
Type: 8 cylinder, 4 cycle, 60 degree vee	
Cooling System: Liquid Ignition: Magneto	
Displacement:	1100 cubic inches
Bore and Stroke:	5.4 x 6 inches
Compression Ratio:	7.5:1
Net Horsepower (max):	450 hp at 2600 rpm
Gross Horsepower (max):	500 hp at 2600 rpm
Net Torque (max):	950 ft-lb at 2200 rpm
Gross Torque (max):	1040 ft-lb at 2200 rpm
Weight:	1414 lb, dry
Fuel: 80 octane gasoline	179 gallons
Engine Oil:	32 quarts

POWER TRAIN
Transmission: Electric drive with speed infinitely variable both forward and reverse
Steering: Electric
Brakes: Electric and mechanical Bendix
Final Drive: Spur gear   Gear Ratio: 5.31:1
Drive Sprocket: At rear of vehicle with 13 teeth
Pitch Diameter: 25.054 inches

RUNNING GEAR
Suspension: Torsion bar
12 individually sprung dual road wheels (6/track)
Tire Size: 26 x 4.3 inches
10 dual track return rollers (5/track)
Dual compensating idler at front of each track
Idler Tire Size: 26 x 4.3 inches
Shock absorbers fitted on first 2 and last 2 road wheels on each side
Tracks: Center guide, cast steel
Type: Single pin, 19 inch width
Pitch: 6 inches
Shoes per Vehicle: 164 (82/track)
Ground Contact Length: 148.3 inches, left side
152.3 inches, right side

ELECTRICAL SYSTEM
Nominal Voltage: 24 volts DC
Battery Charging Generator: 24 volts, 200 amperes, driven by power take-off from main engine
Auxiliary Generator: 24 volts, 50 amperes, driven by the auxiliary engine
Battery: (2) 12 volts in series

COMMUNICATIONS
Radio: SCR 508 or 528 in rear of turret; SCR 506 (command tanks only) on shelf in front of loader
Interphone: (part of radio) 5 stations
Flag Set M238, Panel Set AP50A, Spotlight
Flares: 3 each, M17, M18, M19, and M21 (command tanks only)
Ground Signals Projector M4 (command tanks only)

FIRE AND GAS PROTECTION
(2) 10 pound carbon dioxide, fixed
(2) 4 pound carbon dioxide, portable
(2) 1½ quart decontaminating apparatus

PERFORMANCE

Maximum Speed: Sustained, level road		35 miles/hour
Maximum Tractive Effort: TE at stall		56,000 pounds
Per Cent of Vehicle Weight: TE/W		70.5 per cent
Maximum Grade:		60 per cent
Maximum Trench:		8 feet
Maximum Vertical Wall:		36 inches
Maximum Fording Depth:		48 inches
Minimum Turning Circle: (diameter)		pivot
Cruising Range: Roads	approx.	100 miles

MEDIUM TANK T23 HVSS

GENERAL DATA
Crew: 5 men
Length: Gun forward, over sandshields 308 inches
Length: Gun to rear, over sandshields 264 inches
Length: Over sandshields, w/o gun 237 inches
Gun Overhang: Gun forward 71 inches
Width: Over sandshields 137 inches
Height: To top of cupola 98 inches
Tread: 108 inches
Ground Clearance: 18 inches
Fire Height: 76 inches
Turret Ring Diameter: (inside) 69 inches
Weight, Combat Loaded: 82,000 pounds
Weight, Unstowed: 75,000 pounds
Power to Weight Ratio: Net 11.0 hp/ton
Gross 12.2 hp/ton
Ground Pressure: Zero penetration 11.8 psi

ARMOR
Type: Turret, cast homogeneous steel; Hull, rolled and cast homogeneous steel; Welded assembly

Hull Thickness:	Actual	Angle w/Vertical
Front, Upper	3.0 inches	47 degrees
Lower	2.5 inches	56 degrees
Sides, Front	2.0 inches	0 degrees
Rear	1.5 inches	0 degrees
Rear	1.5 inches	0 to 30 degrees
Top	0.75 inches	90 degrees
Floor, Front	1.0 inches	90 degrees
Rear	0.5 inches	90 degrees

Turret Thickness:		
Gun Shield	3.5 inches	0 degrees
Front	3.0 inches	0 degrees
Sides	2.5 inches	0 to 13 degrees
Rear	2.5 inches	0 degrees
Top	1.0 inches	90 degrees

ARMAMENT
Primary: 76mm Gun M1A1 in Mount M62 (T80) in turret
Traverse: Hydraulic and manual 360 degrees
Traverse Rate: (max) 15 seconds/360 degrees
Elevation: Manual +25 to -10 degrees
Firing Rate: (max) 20 rounds/minute
Loading System: Manual
Stabilizer System: Elevation only
Secondary:
(1) .50 caliber MG HB M2 flexible AA mount on turret
(1) .30 caliber MG M1919A4 coaxial w/76mm gun in turret
(1) .30 caliber MG M1919A4 in bow mount
(1) 2 inch Mortar M3 (smoke) fixed in turret
Provision for (4) .45 caliber SMG M3
Provision for (1) .30 caliber Carbine M1

AMMUNITION
66 rounds 76mm
300 rounds .50 caliber
600 rounds .45 caliber
5000 rounds .30 caliber
12 rounds 2 in. (smoke)
12 hand grenades

FIRE CONTROL AND VISION EQUIPMENT

Primary Weapon:	Direct	Indirect
	Telescope M71D	Azimuth Indicator M19
	Periscope M4A1	Gunner's Quadrant M1
	with Telescope M47	Elevation Quadrant M9

Vision Devices:	Direct	Indirect
Driver	Hatch	Periscope M6 (2)
Asst. Driver	Hatch	Periscope M6 (2)
Commander	Vision cupola and hatch	Periscope M6 (1)
Gunner	None	Periscope M4A1 (1)
Loader	Hatch and pistol port	Periscope M6 (1)

Total Periscopes: M4A1 (1), M6 (6)
Total Pistol Ports: Hull (0), Turret (1)
Vision Cupolas: (1) w/6 vision blocks on turret top

ENGINE
Make and Model: Ford GAN
Type: 8 cylinder, 4 cycle, 60 degree vee
Cooling System: Liquid   Ignition: Magneto
Displacement: 1100 cubic inches
Bore and Stroke: 5.4 x 6 inches
Compression Ratio: 7.5:1
Net Horsepower (max): 450 hp at 2600 rpm
Gross Horsepower (max): 500 hp at 2600 rpm
Net Torque (max): 950 ft-lb at 2200 rpm
Gross Torque (max): 1040 ft-lb at 2200 rpm
Weight: 1414 lb, dry
Fuel: 80 octane gasoline 179 gallons
Engine Oil: 32 quarts

POWER TRAIN
Transmission: Electric drive with speed infinitely variable both forward and reverse
Steering: Electric
Brakes: Electric and mechanical Bendix
Final Drive: Spur gear   Gear Ratio: 5.31:1
Drive Sprocket: At rear of vehicle with 13 teeth
Pitch Diameter: 25.038 inches

RUNNING GEAR
Suspension: Horizontal volute spring
12 dual wheels in 6 bogies (3 bogies/track)
Tire Size: 20 x 6.25 inches
4 dual track return rollers (2/track)
6 single track return rollers (3/track)
Dual adjustable idler at front of each track
Idler Tire Size: 22 x 6.25 inches
Shock absorbers fitted on each bogie
Tracks: Center guide, T80
Type: (T80) Double pin, 23 inch width
Pitch: 6 inches
Shoes per Vehicle: 158 (79/track)
Ground Contact Length: 151 inches

ELECTRICAL SYSTEM
Nominal Voltage: 24 volts DC
Battery Charging Generator: 24 volts, 200 amperes, driven by power take-off from main engine
Auxiliary Generator: 24 volts, 50 amperes, driven by the auxiliary engine
Battery: (2) 12 volts in series

COMMUNICATIONS
Radio: SCR 508 or 528 in rear of turret; SCR 506 (command tanks only) on shelf in front of loader
Interphone: (part of radio) 5 stations
Flag Set M238, Panel Set AP50A, Spotlight
Flares: 3 each, M17, M18, M19, and M21 (command tanks only)
Ground Signals Projector M4 (command tanks only)

FIRE AND GAS PROTECTION
(2) 10 pound carbon dioxide, fixed
(2) 4 pound carbon dioxide, portable
(2) 1½ quart decontaminating apparatus

PERFORMANCE
Maximum Speed: Sustained, level road 35 miles/hour
Maximum Tractive Effort: TE at stall 56,000 pounds
Per Cent of Vehicle Weight: TE/W 68.3 per cent
Maximum Grade: 60 per cent
Maximum Trench: 8 feet
Maximum Vertical Wall: 36 inches
Maximum Fording Depth: 48 inches
Minimum Turning Circle: (diameter) pivot
Cruising Range: Roads approx. 100 miles

GENERAL DATA

Crew:	5 men
Length: Gun forward, over sandshields	312 inches
Length: Gun to rear, over sandshields	270 inches
Length: Over sandshields, w/o gun	237 inches
Gun Overhang: Gun forward	75 inches
Width: Over sandshields	137 inches
Height: To top of cupola	108 inches
Tread:	108 inches
Ground Clearance:	18 inches
Fire Height:	77 inches
Turret Ring Diameter: (inside)	69 inches
Weight, Combat Loaded:	84,210 pounds
Weight, Unstowed:	78,810 pounds
Power to Weight Ratio: Net	10.9 hp/ton
Gross	12.2 hp/ton
Ground Pressure: Zero penetration	12.3 psi

ARMOR

Type: Turret, cast homogeneous steel; Hull, rolled and cast homogeneous steel; Welded assembly

Hull Thickness:	Actual	Angle w/Vertical
Front, Upper	3.0 inches	47 degrees
Lower	2.5 inches	56 degrees
Sides, Front	2.0 inches	0 degrees
Rear	1.5 inches	0 degrees
Rear	1.5 inches	0 to 30 degrees
Top	0.75 inches	90 degrees
Floor, Front	1.0 inches	90 degrees
Rear	0.5 inches	90 degrees
Turret Thickness:		
Gun Shield	3.5 inches	0 degrees
Front	3.0 inches	0 degrees
Sides	2.5 inches	0 to 8 degrees
Rear	2.5 inches	0 to 5 degrees
Top	1.0 inches	90 degrees

ARMAMENT

Primary: 90mm Gun M3 (T7) in Mount T99 in turret

Traverse: Hydraulic and manual	360 degrees
Traverse Rate: (max)	15 seconds/360 degrees
Elevation: Manual	+20 to -10 degrees
Firing Rate: (max)	8 rounds/minute
Loading System:	Manual
Stabilizer System:	None

Secondary:
(1) .50 caliber MG HB M2 flexible AA mount on turret
(1) .30 caliber MG M1919A4 coaxial w/90mm gun in turret
(1) .30 caliber MG M1919A4 in bow mount
(1) 2 inch Mortar M3 (smoke) fixed in turret
Provision for (4) .45 caliber SMG M3
Provision for (1) .30 caliber Carbine M1

AMMUNITION

50 rounds 90mm	12 rounds 2 in. (smoke)
500 rounds .50 caliber	12 hand grenades
720 rounds .45 caliber	
4500 rounds .30 caliber	

FIRE CONTROL AND VISION EQUIPMENT

Primary Weapon:	Direct	Indirect
	Telescope M71C	Azimuth Indicator M19
	Periscope M8A1	Gunner's Quadrant M1
	with Telescope T121	Elevation Quadrant M9
Vision Devices:	Direct	Indirect
Driver	Hatch	Periscope M6 (2)
Asst. Driver	Hatch	Periscope M6 (2)
Commander	Vision cupola and hatch	Periscope M6 (1)
Gunner	None	Periscope M8A1 (1)
Loader	Hatch and pistol port	Periscope M6 (1)

Total Periscopes: M8A1 (1), M6 (6)
Total Pistol Ports: Hull (0), Turret (1)
Vision Cupolas: (1) w/6 vision blocks on turret top

ENGINE

Make and Model: Ford GAN	
Type: 8 cylinder, 4 cycle, 60 degree vee	
Cooling System: Liquid Ignition: Magneto	
Displacement:	1100 cubic inches
Bore and Stroke:	5.4 x 6 inches
Compression Ratio:	7.5:1
Net Horsepower (max):	450 hp at 2600 rpm
Gross Horsepower (max):	500 hp at 2600 rpm
Net Torque (max):	950 ft-lb at 2200 rpm
Gross Torque (max):	1040 ft-lb at 2200 rpm
Weight:	1414 lb, dry
Fuel: 80 octane gasoline	179 gallons
Engine Oil:	32 quarts

POWER TRAIN

Transmission: Electric drive with speed infinitely variable both forward and reverse
Steering: Electric
Brakes: Electric and mechanical Bendix
Final Drive: Spur gear   Gear Ratio: 5.31:1
Drive Sprocket: At rear of vehicle with 13 teeth
Pitch Diameter: 25.038 inches

RUNNING GEAR

Suspension: Horizontal volute spring
12 dual wheels in 6 bogies (3 bogies/track)
Tire Size: 20 x 6.25 inches
4 dual track return rollers (2/track)
6 single track return rollers (3/track)
Dual adjustable idler at front of each track
Idler Tire Size: 22 x 6.25 inches
Shock absorbers fitted on each bogie
Tracks: Center guide, T66 and T80
Type: (T66) Single pin, 23 inch width
(T80) Double pin, 23 inch width
Pitch: 6 inches
Shoes per Vehicle: 158 (79/track)
Ground Contact Length: 149 inches

ELECTRICAL SYSTEM

Nominal Voltage: 24 volts DC
Battery Charging Generator: 24 volts, 200 amperes, driven by power take-off from main engine
Auxiliary Generator: 24 volts, 50 amperes, driven by the auxiliary engine
Battery: (2) 12 volts in series

COMMUNICATIONS

Radio: SCR 508 or 528 in rear of turret; SCR 506 (command tanks only) on shelf in front of loader
Interphone: (part of radio) 5 stations
Flag Set M238, Panel Set AP50A, Spotlight
Flares: 3 each, M17, M18, M19, and M21 (command tanks only)
Ground Signals Projector M4 (command tanks only)

FIRE AND GAS PROTECTION

(2) 10 pound carbon dioxide, fixed
(2) 4 pound carbon dioxide, portable
(2) 1½ quart decontaminating apparatus

PERFORMANCE

Maximum Speed: Sustained, level road		30 miles/hour
Short periods, level		34.5 miles/hour
Maximum Tractive Effort: TE at stall		56,000 pounds
Per Cent of Vehicle Weight: TE/W		66.5 per cent
Maximum Grade:		60 per cent
Maximum Trench:		8 feet
Maximum Vertical Wall:		36 inches
Maximum Fording Depth:		48 inches
Minimum Turning Circle: (diameter)		pivot
Cruising Range: Roads	approx.	100 miles

212

MEDIUM TANK T25E1

GENERAL DATA

Crew:	5	men
Length: Gun forward, over pintle	323.3	inches
Length: Gun to rear, over sandshields	268.6	inches
Length: Over ss and pintle, w/o gun	247.8	inches
Gun Overhang: Gun forward	75.5	inches
Width: Over sandshields	127.0	inches
Height: To top of cupola	109.4	inches
Tread:	105	inches
Ground Clearance:	17.2	inches
Fire Height:	78	inches
Turret Ring Diameter: (inside)	69	inches
Weight, Combat Loaded:	77,590	pounds
Weight, Unstowed:	70,000	pounds
Power to Weight Ratio: Net	11.6	hp/ton
Gross	12.9	hp/ton
Ground Pressure: Zero penetration	13.6	psi

ARMOR

Type: Turret, cast homogeneous steel; Hull, rolled and cast homogeneous steel; Welded assembly

Hull Thickness:	Actual		Angle w/Vertical
Front, Upper	3.0	inches	46 degrees
Lower	2.5	inches	53 degrees
Sides, Front	2.0	inches	0 degrees
Rear	1.5	inches	0 degrees
Rear, Upper	1.5	inches	10 degrees
Lower	0.75	inches	62 degrees
Top	0.875	inches	90 degrees
Floor, Front	1.0	inches	90 degrees
Rear	0.5	inches	90 degrees
Turret Thickness:			
Gun Shield	3.5	inches	0 degrees
Front	3.0	inches	0 degrees
Sides	2.5	inches	0 to 8 degrees
Rear	2.5	inches	0 to 5 degrees
Top	1.0	inches	90 degrees

ARMAMENT

Primary: 90mm Gun M3 (T7) in Mount T99 in turret

Traverse: Hydraulic and manual	360 degrees
Traverse Rate: (max)	15 seconds/360 degrees
Elevation: Manual	+20 to -10 degrees
Firing Rate: (max)	8 rounds/minute
Loading System:	Manual
Stabilizer System:	None

Secondary:
- (1) .50 caliber MG HB M2 flexible AA mount on turret
- (1) .30 caliber MG M1919A4 coaxial w/90mm gun in turret
- (1) .30 caliber MG M1919A4 in bow mount
- (1) 2 inch Mortar M3 (smoke) fixed in turret
- Provision for (5) .45 caliber SMG M3

AMMUNITION

42 rounds 90mm	12 rounds 2 in. (smoke)
500 rounds .50 caliber	12 hand grenades
600 rounds .45 caliber	
5000 rounds .30 caliber	

FIRE CONTROL AND VISION EQUIPMENT

Primary Weapon:	Direct	Indirect
	Telescope M71C	Azimuth Indicator M19
	Periscope M8A1	Gunner's Quadrant M1
	with Telescope T121	Elevation Quadrant M9
Vision Devices:	Direct	Indirect
Driver	Hatch	Periscope M6 (2)
Asst. Driver	Hatch	Periscope M6 (2)
Commander	Vision cupola and hatch	Periscope M6 (1)
Gunner	None	Periscope M8A1 (1)
Loader	Hatch	Periscope M6 (1)

Total Periscopes: M8A1 (1), M6 (6)
Vision Cupolas: (1) w/6 vision blocks on turret top

ENGINE

Make and Model: Ford GAF	
Type: 8 cylinder, 4 cycle, 60 degree vee	
Cooling System: Liquid Ignition: Magneto	
Displacement:	1100 cubic inches
Bore and Stroke:	5.4 x 6 inches
Compression Ratio:	7.5:1
Net Horsepower (max):	450 hp at 2600 rpm
Gross Horsepower (max):	500 hp at 2600 rpm
Net Torque (max):	950 ft-lb at 2200 rpm
Gross Torque (max):	1040 ft-lb at 2200 rpm
Weight:	1414 lb, dry
Fuel: 80 octane gasoline	202 gallons
Engine Oil:	32 quarts

POWER TRAIN

Transfer Case: Planetary reduction gears
  Gear Ratio: 1.377:1 engine to transmission
Transmission: Torqmatic, 3 speeds forward, 1 reverse
  Torque Converter Ratio: Varies from 1:1 to 4.8:1

Gear Ratios:	1st	1:1	3rd	0.244:1
	2nd	0.428:1	reverse	0.756:1

Steering: Controlled differential
  Bevel Gear Ratio: 3.53:1   Steering Ratio: 2.08:1
Brakes: Mechanical, external contracting
Final Drive: Spur gear   Gear Ratio: 3.48:1
Drive Sprocket: At rear of vehicle with 13 teeth
  Pitch Diameter: 25.068 inches

RUNNING GEAR

Suspension: Torsion bar
  12 individually sprung dual road wheels (6/track)
  Tire Size: 26 x 4.3 inches
  10 dual track return rollers (5/track)
  Dual compensating idler at front of each track
  Idler Tire Size: 26 x 4.3 inches
  Shock absorbers fitted on first 2 and last 2 road wheels on each side
Tracks: Center guide, cast steel
  Type: Single pin, 19 inch width
  Pitch: 6 inches
  Shoes per Vehicle: 164 (82/track)
  Ground Contact Length: 148.3 inches, left side
                         152.3 inches, right side

ELECTRICAL SYSTEM

Nominal Voltage: 24 volts DC
Generator: (1) 24 volts, 150 amperes, belt driven by either the main engine or the auxiliary engine
Battery: (2) 12 volts in series

COMMUNICATIONS

Radio: SCR 610 in rear of turret
Interphone: RC 99, 4 stations
Flag Set M238, Panel Set AP50A, Spotlight

FIRE AND GAS PROTECTION

- (2) 10 pound carbon dioxide, fixed
- (2) 4 pound carbon dioxide, portable
- (2) 1½ quart decontaminating apparatus

PERFORMANCE

Maximum Speed: Sustained, level road	30 miles/hour
Short periods, level	35 miles/hour
Maximum Tractive Effort: TE at stall	53,000 pounds
Per Cent of Vehicle Weight: TE/W	68.3 per cent
Maximum Grade:	60 per cent
Maximum Trench:	8 feet
Maximum Vertical Wall:	46 inches
Maximum Fording Depth:	48 inches
Minimum Turning Circle: (diameter)	40 feet
Cruising Range: Roads approx.	110 miles

GENERAL DATA

Crew:	5	men
Length: Gun forward, over sandshields	318.5	inches
Length: Gun to rear, over sandshields	270.0	inches
Length: Over sandshields, w/o gun	244.4	inches
Gun Overhang: Gun forward	74.1	inches
Width: Over sandshields	138	inches
Height: To top of cupola	109	inches
Tread:	110	inches
Ground Clearance:	17.2	inches
Fire Height:	78	inches
Turret Ring Diameter: (inside)	69	inches
Weight, Combat Loaded:	95,100	pounds
Weight, Unstowed:	89,700	pounds
Power to Weight Ratio: Net	9.5	hp/ton
Gross	10.5	hp/ton
Ground Pressure: Zero penetration	12.9	psi

ARMOR

Type: Turret, cast homogeneous steel; Hull, rolled and cast homogeneous steel; Welded assembly

Hull Thickness:

		Actual		Angle w/Vertical
Front,	Upper	4.0	inches	46 degrees
	Lower	3.0	inches	53 degrees
Sides,	Front	3.0	inches	0 degrees
	Rear	2.0	inches	0 degrees
Rear		2.0	inches	0 to 30 degrees
Top		0.875	inches	90 degrees
Floor,	Front	1.0	inches	90 degrees
	Rear	0.5	inches	90 degrees

Turret Thickness:

	Actual		Angle w/Vertical
Gun Shield	4.5	inches	0 degrees
Front	4.0	inches	0 degrees
Sides	3.0	inches	0 to 8 degrees
Rear	3.0	inches	0 to 5 degrees
Top	1.0	inches	90 degrees

ARMAMENT

Primary: 90mm Gun M3 (T7) in Mount T99E1 in turret

Traverse: Hydraulic and manual	360 degrees
Traverse Rate: (max)	15 seconds/360 degrees
Elevation: Manual	+20 to -10 degrees
Firing Rate: (max)	8 rounds/minute
Loading System:	Manual
Stabilizer System:	None

Secondary:
- (1) .50 caliber MG HB M2 flexible AA mount on turret
- (1) .30 caliber MG M1919A4 coaxial w/90mm gun in turret
- (1) .30 caliber MG M1919A4 in bow mount
- (1) 2 inch Mortar M3 (smoke) fixed in turret
- Provision for (4) .45 caliber SMG M3
- Provision for (1) .30 caliber Carbine M1

AMMUNITION

58 rounds 90mm	12 rounds 2 in. (smoke)
500 rounds .50 caliber	12 hand grenades
720 rounds .45 caliber	
4500 rounds .30 caliber	

FIRE CONTROL AND VISION EQUIPMENT

Primary Weapon:	Direct	Indirect
	Telescope M71C	Azimuth Indicator M19
	Periscope M8A1	Gunner's Quadrant M1
	with Telescope T121	Elevation Quadrant M9

Vision Devices:	Direct	Indirect
Driver	Hatch	Periscope M6 (2)
Asst. Driver	Hatch	Periscope M6 (2)
Commander	Vision cupola and hatch	Periscope M6 (1)
Gunner	None	Periscope M8A1 (1)
Loader	Hatch and pistol port	Periscope M6 (1)

Total Periscopes: M8A1 (1), M6 (6)
Total Pistol Ports: Hull (0), Turret (1)
Vision Cupolas: (1) w/6 vision blocks on turret top

ENGINE

Make and Model: Ford GAN	
Type: 8 cylinder, 4 cycle, 60 degree vee	
Cooling System: Liquid Ignition: Magneto	
Displacement:	1100 cubic inches
Bore and Stroke:	5.4 x 6 inches
Compression Ratio:	7.5:1
Net Horsepower (max):	450 hp at 2600 rpm
Gross Horsepower (max):	500 hp at 2600 rpm
Net Torque (max):	950 ft-lb at 2200 rpm
Gross Torque (max):	1040 ft-lb at 2200 rpm
Weight:	1414 lb, dry
Fuel: 80 octane gasoline	179 gallons
Engine Oil:	32 quarts

POWER TRAIN

Transmission: Electric drive with speed infinitely variable both forward and reverse
Steering: Electric
Brakes: Electric and mechanical Bendix
Final Drive: Spur gear   Gear Ratio: 6.5:1
Drive Sprocket: At rear of vehicle with 13 teeth
    Pitch Diameter: 25.068 inches

RUNNING GEAR

Suspension: Torsion bar
    12 individually sprung dual road wheels (6/track)
    Tire Size: 26 x 6 inches
    10 dual track return rollers (5/track)
    Dual compensating idler at front of each track
    Idler Tire Size: 26 x 6 inches
    Shock absorbers fitted on first 2 and last 2 road wheels on each side
Tracks: Center guide, T81 and T80E1
    Type: (T81) Single pin, 24 inch width
        (T80E1) Double pin, 23 inch width
    Pitch: 6 inches
    Shoes per Vehicle: 164 (82/track)
    Ground Contact Length: 151.5 inches, left side
                           155.5 inches, right side

ELECTRICAL SYSTEM

Nominal Voltage: 24 volts DC
Battery Charging Generator: 24 volts, 200 amperes, driven by power take-off from main engine
Auxiliary Generator: 24 volts, 50 amperes, driven by the auxiliary engine
Battery: (2) 12 volts in series

COMMUNICATIONS

Radio: SCR 508 or 528 in rear of turret; SCR 506 (command tanks only) on shelf in front of loader
Interphone: (part of radio) 5 stations
Flag Set M238, Panel Set AP50A, Spotlight
Flares: 3 each, M17, M18, M19, and M21 (command tanks only)
Ground Signals Projector M4 (command tanks only)

FIRE AND GAS PROTECTION

(2) 10 pound carbon dioxide, fixed
(2) 4 pound carbon dioxide, portable
(2) 1½ quart decontaminating apparatus

PERFORMANCE

Maximum Speed: Sustained, level road	25 miles/hour
Short periods, level	28 miles/hour
Maximum Tractive Effort: TE at stall	69,000 pounds
Per Cent of Vehicle Weight: TE/W	72.6 per cent
Maximum Grade:	60 per cent
Maximum Trench:	8 feet
Maximum Vertical Wall:	36 inches
Maximum Fording Depth:	48 inches
Minimum Turning Circle: (diameter)	pivot
Cruising Range: Roads approx.	75 miles

MEDIUM TANK T26E1

GENERAL DATA

Crew:	5 men
Length: Gun forward, over pintle	324.6 inches
Length: Gun to rear, over sandshields	269 inches
Length: Over ss and pintle, w/o gun	249.1 inches
Gun Overhang: Gun forward	75.5 inches
Width: Over sandshields	138.3 inches
Height: To top of cupola	109.4 inches
Tread:	110 inches
Ground Clearance:	17.2 inches
Fire Height:	78 inches
Turret Ring Diameter: (inside)	69 inches
Weight, Combat Loaded:	87,350 pounds
Weight, Unstowed:	81,300 pounds
Power to Weight Ratio: Net	10.3 hp/ton
Gross	11.4 hp/ton
Ground Pressure: Zero penetration	12.1 psi

ARMOR

Type: Turret, cast homogeneous steel; Hull, rolled and cast homogeneous steel; Welded assembly

Hull Thickness:		Actual		Angle w/Vertical
Front,	Upper	4.0	inches	46 degrees
	Lower	3.0	inches	53 degrees
Sides,	Front	3.0	inches	0 degrees
	Rear	2.0	inches	0 degrees
Rear,	Upper	2.0	inches	10 degrees
	Lower	0.75	inches	62 degrees
Top		0.875	inches	90 degrees
Floor,	Front	1.0	inches	90 degrees
	Rear	0.5	inches	90 degrees

Turret Thickness:			
Gun Shield	4.5	inches	0 degrees
Front	4.0	inches	0 degrees
Sides	3.0	inches	0 to 8 degrees
Rear	3.0	inches	0 to 5 degrees
Top	1.0	inches	90 degrees

ARMAMENT

Primary: 90mm Gun M3 (T7) in Mount T99E1 in turret

Traverse: Hydraulic and manual	360 degrees
Traverse Rate: (max)	15 seconds/360 degrees
Elevation: Manual	+20 to -10 degrees
Firing Rate: (max)	8 rounds/minute
Loading System:	Manual
Stabilizer System:	None

Secondary:
(1) .50 caliber MG HB M2 flexible AA mount on turret
(1) .30 caliber MG M1919A4 coaxial w/90mm gun in turret
(1) .30 caliber MG M1919A4 in bow mount
(1) 2 inch Mortar M3 (smoke) fixed in turret
Provision for (5) .45 caliber SMG M3

AMMUNITION

42 rounds 90mm	12 rounds 2 in. (smoke)
500 rounds .50 caliber	12 hand grenades
600 rounds .45 caliber	
5000 rounds .30 caliber	

FIRE CONTROL AND VISION EQUIPMENT

Primary Weapon:	Direct	Indirect
	Telescope M71C	Azimuth Indicator M19
	Periscope M8A1	Gunner's Quadrant M1
	with Telescope T121	Elevation Quadrant M9

Vision Devices:	Direct	Indirect
Driver	Hatch	Periscope M6 (2)
Asst. Driver	Hatch	Periscope M6 (2)
Commander	Vision cupola and hatch	Periscope M6 (1)
Gunner	None	Periscope M8A1 (1)
Loader	Hatch	Periscope M6 (1)

Total Periscopes: M8A1 (1), M6 (6)
Vision Cupolas: (1) w/6 vision blocks on turret top

ENGINE

Make and Model: Ford GAF	
Type: 8 cylinder, 4 cycle, 60 degree vee	
Cooling System: Liquid Ignition: Magneto	
Displacement:	1100 cubic inches
Bore and Stroke:	5.4 x 6 inches
Compression Ratio:	7.5:1
Net Horsepower (max):	450 hp at 2600 rpm
Gross Horsepower (max):	500 hp at 2600 rpm
Net Torque (max):	950 ft-lb at 2200 rpm
Gross Torque (max):	1040 ft-lb at 2200 rpm
Weight:	1414 lb, dry
Fuel: 80 octane gasoline	202 gallons
Engine Oil:	32 quarts

POWER TRAIN

Transfer Case: Planetary reduction gears
    Gear Ratio: 1.377:1 engine to transmission
Transmission: Torqmatic, 3 speeds forward, 1 reverse
    Torque Converter Ratio: Varies from 1:1 to 4.8:1

Gear Ratios:	1st	1:1	3rd	0.244:1
	2nd	0.428:1	reverse	0.756:1

Steering: Controlled differential
    Bevel Gear Ratio: 3.53:1   Steering Ratio: 2.08:1
Brakes: Mechanical, external contracting
Final Drive: Spur gear   Gear Ratio: 3.95:1
Drive Sprocket: At rear of vehicle with 13 teeth
    Pitch Diameter: 25.068 inches

RUNNING GEAR

Suspension: Torsion bar
    12 individually sprung dual road wheels (6/track)
    Tire Size: 26 x 6 inches
    10 dual track return rollers (5/track)
    Dual compensating idler at front of each track
    Idler Tire Size: 26 x 6 inches
    Shock absorbers fitted on first 2 and last 2 road wheels on each side
Tracks: Center guide, T81 cast steel
    Type: (T81) Single pin, 24 inch width
    Pitch: 6 inches
    Shoes per Vehicle: 164 (82/track)
    Ground Contact Length: 148.3 inches, left side
                           152.3 inches, right side

ELECTRICAL SYSTEM

Nominal Voltage: 24 volts DC
Generator: (1) 24 volts, 150 amperes, belt driven by either the main engine or the auxiliary engine
Battery: (2) 12 volts in series

COMMUNICATIONS

Radio: SCR 610 in rear of turret
Interphone: RC 99, 4 stations
Flag Set M238, Panel Set AP50A, Spotlight

FIRE AND GAS PROTECTION

(2) 10 pound carbon dioxide, fixed
(2) 4 pound carbon dioxide, portable
(2) 1½ quart decontaminating apparatus

PERFORMANCE

Maximum Speed: Sustained, level road	25 miles/hour
Short periods, level	30 miles/hour
Maximum Tractive Effort: TE at stall	60,000 pounds
Per Cent of Vehicle Weight: TE/W	68.7 per cent
Maximum Grade:	60 per cent
Maximum Trench:	8 feet
Maximum Vertical Wall:	46 inches
Maximum Fording Depth:	48 inches
Minimum Turning Circle: (diameter)	40 feet
Cruising Range: Roads approx.	100 miles

MEDIUM TANK M45 (T26E2)

GENERAL DATA
Crew: 5 men
Length: Howitzer forward, over muffler — 256.7 inches
Length: Howitzer to rear, over muffler and ss — 254.3 inches
Length: over ss and muffler, w/o howitzer — 254.3 inches
Gun Overhang: Howitzer forward — 2.4 inches
Width: Over sandshields — 138.3 inches
Height: Over mg Mount — 110.9 inches
Tread: — 110 inches
Ground Clearance: — 18.8 inches
Fire Height: — 80 inches
Turret Ring Diameter: (inside) — 69 inches
Weight, Combat Loaded: — 93,000 pounds
Weight, Unstowed: — 86,000 pounds
Power to Weight Ratio: Net — 9.7 hp/ton
           Gross — 10.8 hp/ton
Ground Pressure: Zero penetration — 13.2 psi

ARMOR
Type: Turret, cast homogeneous steel; Hull, rolled and cast homogeneous steel; Welded assembly

Hull Thickness:

		Actual		Angle w/Vertical
Front,	Upper	4.0	inches	46 degrees
	Lower	3.0	inches	53 degrees
Sides,	Front	3.0	inches	0 degrees
	Rear	2.0	inches	0 degrees
Rear,	Upper	2.0	inches	10 degrees
	Lower	0.75	inches	62 degrees
Top		0.875	inches	90 degrees
Floor,	Front	1.0	inches	90 degrees
	Rear	0.5	inches	90 degrees

Turret Thickness:

Howitzer Shield	8.0	inches		0 degrees
Front	5.0	inches		0 degrees
Sides	5.0 to 3.0	inches		0 to 8 degrees
Rear	2.5	inches		0 to 5 degrees
Top	1.0	inches		90 degrees

ARMAMENT
Primary: 105mm Howitzer M4 in Mount M71 (T117) in turret
Traverse: Hydraulic and manual — 360 degrees
Traverse Rate: (max) — 15 seconds/360 degrees
Elevation: Manual — +35 to -10 degrees
Firing Rate: (max) — 8 rounds/minute
Loading System: — Manual
Stabilizer System: — Elevation only

Secondary:
(1) .50 caliber MG HB M2 flexible AA mount on turret
(1) .30 caliber MG M1919A4 coaxial w/105mm howitzer in turret
(1) .30 caliber MG M1919A4 in bow mount
Provision for (5) .45 caliber SMG M3
Provision for (1) .30 caliber Carbine M2 w/grenade launcher

AMMUNITION
74 rounds 105mm   12 hand grenades
550 rounds .50 caliber
900 rounds .45 caliber
5000 rounds .30 caliber

FIRE CONTROL AND VISION EQUIPMENT*

Primary Weapon:	Direct	Indirect
	Telescope M76G	Azimuth Indicator M20
	Periscope M10D	Elevation Quadrant M9
	Elbow Telescope M62	

Vision Devices:	Direct	Indirect
Driver	Hatch	Periscope M13 (1)
Asst. Driver	Hatch	Periscope M13 (1)
Commander	Vision cupola and hatch	Periscope M15 (1)
Gunner	None	Periscope M10D (1)
Loader	Hatch and pistol port	Periscope M13 (1)

Total Periscopes: M10D (1), M13 (3), M15 (1); In addition the following spares are carried, M10D (1), M13 (4)
Total Pistol Ports: Hull (0), Turret (1)
Vision Cupolas: (1) w/6 vision blocks on turret top

*Late production vision devices, drivers' auxiliary periscopes eliminated

ENGINE
Make and Model: Ford GAF
Type: 8 cylinder, 4 cycle, 60 degree vee
Cooling System: Liquid   Ignition: Magneto
Displacement: — 1100 cubic inches
Bore and Stroke: — 5.4 x 6 inches
Compression Ratio: — 7.5:1
Net Horsepower (max): — 450 hp at 2600 rpm
Gross Horsepower (max): — 500 hp at 2600 rpm
Net Torque (max): — 950 ft-lb at 2200 rpm
Gross Torque (max): — 1040 ft-lb at 2200 rpm
Weight: — 1414 lb, dry
Fuel: 80 octane gasoline — 183 gallons
Engine Oil: — 32 quarts

POWER TRAIN
Transfer Case: Planetary reduction gears
   Gear Ratio: 1.377:1 engine to transmission
Transmission: Torqmatic, 3 speeds forward, 1 reverse
   Torque Converter Ratio: Varies from 1:1 to 4.8:1

Gear Ratios:	1st	1:1	3rd	0.244:1
	2nd	0.428:1	reverse	0.756:1

Steering: Controlled differential
Bevel Gear Ratio: 3.53:1   Steering Ratio: 1.79:1
Brakes: Mechanical, 3 shoe, reverse anchor
Final Drive: Spur gear   Gear Ratio: 3.95:1
Drive Sprocket: At rear of vehicle with 13 teeth
   Pitch Diameter: 25.068 inches

RUNNING GEAR
Suspension: Torsion bar
   12 individually sprung dual road wheels (6/track)
   Tire Size: 26 x 6 inches
   10 dual track return rollers (5/track)
   Dual compensating idler at front of each track
   Idler Tire Size: 26 x 6 inches
   Shock absorbers fitted on first 2 and last 2 road wheels on each side
Tracks: Center guide, T81 and T80E1
   Type: (T81) Single pin, 24 inch width
       (T80E1) Double pin, 23 inch width
   Pitch: 6 inches
   Shoes per Vehicle: 164 (82/track)
   Ground Contact Length: 151.5 inches, left side
                   155.5 inches, right side

ELECTRICAL SYSTEM
Nominal Voltage: 24 volts DC
Generator: (1) 24 volts, 150 amperes, belt driven by either the main engine or the auxiliary engine
Battery: (2) 12 volts in series

COMMUNICATIONS
Radio: SCR 508 or 528 in rear of turret; SCR 506 (command tanks only) on shelf in front of loader; AN/VRC 3 in rear of turret on some tanks (infantry communication)
Interphone: (part of radio) 5 stations
Flag Set M238, Panel Set AP50A, Spotlight
Flares: 3 each, M17, M18, M19, and M21 (command tanks only)
Ground Signals Projector M4 (command tanks only)

FIRE AND GAS PROTECTION
(2) 10 pound carbon dioxide, fixed
(2) 4 pound carbon dioxide, portable
(2) 1½ quart decontaminating apparatus

PERFORMANCE
Maximum Speed: Sustained, level road — 25 miles/hour
          Short periods, level — 30 miles/hour
Maximum Tractive Effort: TE at stall — 60,000 pounds
   Per Cent of Vehicle Weight: TE/W — 64.5 per cent
Maximum Grade: — 60 per cent
Maximum Trench: — 8 feet
Maximum Vertical Wall: — 46 inches
Maximum Fording Depth: — 48 inches
Minimum Turning Circle: (diameter) — 60 feet
Cruising Range: Roads — approx. 100 miles

MEDIUM TANK M26 (T26E3)

GENERAL DATA

Crew:	5	men
Length: Gun forward, over pintle	340.5	inches
Length: Gun to rear, over sandshields	285	inches
Length: Over ss and pintle, w/o gun	249.1	inches
Gun Overhang: Gun forward	91.4	inches
Width: Over sandshields	138.3	inches
Height: To top of cupola	109.4	inches
Tread:	110	inches
Ground Clearance:	17.2	inches
Fire Height:	78	inches
Turret Ring Diameter: (inside)	69	inches
Weight, Combat Loaded:	92,355	pounds
Weight, Unstowed:	84,850	pounds
Power to Weight Ratio: Net	9.7	hp/ton
Gross	10.8	hp/ton
Ground Pressure: Zero penetration	12.5	psi

ARMOR

Type: Turret, cast homogeneous steel; Hull, rolled and cast homogeneous steel; Welded assembly

Hull Thickness:

	Actual		Angle w/Vertical
Front, Upper	4.0	inches	46 degrees
Lower	3.0	inches	53 degrees
Sides, Front	3.0	inches	0 degrees
Rear	2.0	inches	0 degrees
Rear, Upper	2.0	inches	10 degrees
Lower	0.75	inches	62 degrees
Top	0.875	inches	90 degrees
Floor, Front	1.0	inches	90 degrees
Rear	0.5	inches	90 degrees

Turret Thickness:

Gun Shield	4.5	inches	0 degrees
Front	4.0	inches	0 degrees
Sides	3.0	inches	0 to 8 degrees
Rear	3.0	inches	0 to 5 degrees
Top	1.0	inches	90 degrees

ARMAMENT

Primary: 90mm Gun M3 (T7) in Mount M67 (T99E2) in turret

Traverse: Hydraulic and manual	360 degrees
Traverse Rate: (max)	15 seconds/360 degrees
Elevation: Manual	+20 to -10 degrees
Firing Rate: (max)	8 rounds/minute
Loading System:	Manual
Stabilizer System:	None

Secondary:
- (1) .50 caliber MG HB M2 flexible AA mount on turret
- (1) .30 caliber MG M1919A4 coaxial w/90mm gun in turret
- (1) .30 caliber MG M1919A4 in bow mount
- Provision for (5) .45 caliber SMG M3
- Provision for (1) .30 caliber Carbine M2 w/grenade launcher

AMMUNITION

70 rounds 90mm	12 hand grenades
550 rounds .50 caliber	
900 rounds .45 caliber	
5000 rounds .30 caliber	

FIRE CONTROL AND VISION EQUIPMENT

Primary Weapon:	Direct	Indirect
	Telescope M71C	Azimuth Indicator M19
	Periscope M10F	Gunner's Quadrant M1
	Periscope M4A1, spare	Elevation Quadrant M9

Vision Devices:	Direct	Indirect
Driver	Hatch	Periscope M6 (2)
Asst. Driver	Hatch	Periscope M6 (2)
Commander	Vision cupola and hatch	Periscope M6 (1)
Gunner	None	Periscope M10F (1)
Loader	Hatch and pistol port	Periscope M6 (1)

Total Periscopes: M10F (1), M6 (6)
Total Pistol Ports: Hull (0), Turret (1)
Vision Cupolas: (1) w/6 vision blocks on turret top

ENGINE

Make and Model: Ford GAF	
Type: 8 cylinder, 4 cycle, 60 degree vee	
Cooling System: Liquid Ignition: Magneto	
Displacement:	1100 cubic inches
Bore and Stroke:	5.4 x 6 inches
Compression Ratio:	7.5:1
Net Horsepower (max):	450 hp at 2600 rpm
Gross Horsepower (max):	500 hp at 2600 rpm
Net Torque (max):	950 ft-lb at 2200 rpm
Gross Torque (max):	1040 ft-lb at 2200 rpm
Weight:	1414 lb, dry
Fuel: 80 octane gasoline	183 gallons
Engine Oil:	32 quarts

POWER TRAIN

Transfer Case: Planetary reduction gears
  Gear Ratio: 1.377:1 engine to transmission
Transmission: Torqmatic, 3 speeds forward, 1 reverse
  Torque Converter Ratio: Varies from 1:1 to 4.8:1

Gear Ratios:	1st	1:1	3rd	0.244:1
	2nd	0.428:1	reverse	0.756:1

Steering: Controlled differential
  Bevel Gear Ratio: 3.53:1   Steering Ratio: 1.79:1
Brakes: Mechanical, 3 shoe, reverse anchor
Final Drive: Spur gear   Gear Ratio: 3.95:1
Drive Sprocket: At rear of vehicle with 13 teeth
  Pitch Diameter: 25.068 inches

RUNNING GEAR

Suspension: Torsion bar
  12 individually sprung dual road wheels (6/track)
  Tire Size: 26 x 6 inches
  10 dual track return rollers (5/track)
  Dual compensating idler at front of each track
  Idler Tire Size: 26 x 6 inches
  Shock absorbers fitted on first 2 and last 2 road wheels on each side
Tracks: Center guide, T81 and T80E1
  Type: (T81) Single pin, 24 inch width
        (T80E1) Double pin, 23 inch width
  Pitch: 6 inches
  Shoes per Vehicle: 164 (82/track)
  Ground Contact Length: 151.5 inches, left side
                         155.5 inches, right side

ELECTRICAL SYSTEM

Nominal Voltage: 24 volts DC
Generator: (1) 24 volts, 150 amperes, belt driven by either the main engine or the auxiliary engine
Battery: (2) 12 volts in series

COMMUNICATIONS

Radio: SCR 508 or 528 in rear of turret; SCR 506 (command tanks only) on shelf in front of loader; AN/VRC 3 in rear of turret on some tanks (infantry communication)
Interphone: (part of radio) 5 stations
Flag Set M238, Panel Set AP50A, Spotlight
Flares: 3 each, M17, M18, M19, and M21 (command tanks only)
Ground Signals Projector M4 (command tanks only)

FIRE AND GAS PROTECTION

- (2) 10 pound carbon dioxide, fixed
- (2) 4 pound carbon dioxide, portable
- (2) 1½ quart decontaminating apparatus

PERFORMANCE

Maximum Speed: Sustained, level road	25 miles/hour
Short periods, level	30 miles/hour
Maximum Tractive Effort: TE at stall	60,000 pounds
Per Cent of Vehicle Weight: TE/W	65.0 per cent
Maximum Grade:	60 per cent
Maximum Trench:	8 feet
Maximum Vertical Wall:	46 inches
Maximum Fording Depth:	48 inches
Minimum Turning Circle: (diameter)	60 feet
Cruising Range: Roads approx.	100 miles

MEDIUM TANK T26E4

GENERAL DATA

Crew:	5	men
Length: Gun forward, over pintle	406	inches
Length: Gun to rear, over sandshields	348	inches
Length: Over ss and pintle, w/o gun	249.1	inches
Gun Overhang: Gun forward	157	inches
Width: Over sandshields	138.3	inches
Height: To top of cupola	109.4	inches
Tread:	110	inches
Ground Clearance:	17.2	inches
Fire Height:	78	inches
Turret Ring Diameter: (inside)	69	inches
Weight, Combat Loaded:	96,000	pounds
Weight, Unstowed:	88,000	pounds
Power to Weight Ratio: Net	9.4	hp/ton
Gross	10.4	hp/ton
Ground Pressure: Zero penetration	13.0	psi

ARMOR

Type: Turret, cast homogeneous steel; Hull, rolled and cast homogeneous steel; Welded assembly

Hull Thickness:

		Actual	Angle w/Vertical
Front, Upper	4.0	inches	46 degrees
Lower	3.0	inches	53 degrees
Sides, Front	3.0	inches	0 degrees
Rear	2.0	inches	0 degrees
Rear, Upper	2.0	inches	10 degrees
Lower	0.75	inches	62 degrees
Top	0.875	inches	90 degrees
Floor, Front	1.0	inches	90 degrees
Rear	0.5	inches	90 degrees

Turret Thickness:

Gun Shield	4.5	inches	0 degrees
Front	4.0	inches	0 degrees
Sides	3.0	inches	0 to 8 degrees
Rear	3.0 inches plus counterweight		0 to 5 degrees
Top	1.0	inches	90 degrees

ARMAMENT

Primary: 90 mm Gun T15E2 in Mount T119 in turret

Traverse: Hydraulic and manual	360 degrees
Traverse Rate: (max)	15 seconds/360 degrees
Elevation: Manual	+20 to -10 degrees
Firing Rate: (max)	4 rounds/minute
Loading System:	Manual
Stabilizer System:	None

Secondary:
(1) .50 caliber MG HB M2 flexible AA mount on turret
(1) .30 caliber MG M1919A4 coaxial w/90mm gun in turret
(1) .30 caliber MG M1919A4 in bow mount
Provision for (5) .45 caliber SMG M3
Provision for (1) .30 caliber Carbine M2 w/grenade launcher

AMMUNITION

54 rounds 90mm	12 hand grenades
440 rounds .50 caliber	
900 rounds .45 caliber	
5000 rounds .30 caliber	

FIRE CONTROL AND VISION EQUIPMENT*

Primary Weapon:	Direct	Indirect
	Telescope M71E4	Azimuth Indicator M200
	Periscope M10E4	Elevation Quadrant M9

Vision Devices:	Direct	Indirect
Driver	Hatch	Periscope M13 (1)
Asst. Driver	Hatch	Periscope M13 (1)
Commander	Vision cupola and hatch	Periscope M15 (1)
Gunner	None	Periscope M10E4 (1)
Loader	Hatch and pistol port	Periscope M13 (1)

Total Periscopes: M10E4 (1), M13 (3), M15 (1); In addition the following spares are carried, M10E4 (1), M13 (4)
Total Pistol Ports: Hull (0), Turret (1)
Vision Cupolas: (1) w/6 vision blocks on turret top

*Late production vision devices, drivers' auxiliary periscopes eliminated

ENGINE

Make and Model: Ford GAF	
Type: 8 cylinder, 4 cycle, 60 degree vee	
Cooling System: Liquid Ignition: Magneto	
Displacement:	1100 cubic inches
Bore and Stroke:	5.4 x 6 inches
Compression Ratio:	7.5:1
Net Horsepower (max):	450 hp at 2600 rpm
Gross Horsepower (max):	500 hp at 2600 rpm
Net Torque (max):	950 ft-lb at 2200 rpm
Gross Torque (max):	1040 ft-lb at 2200 rpm
Weight:	1414 lb, dry
Fuel: 80 octane gasoline	183 gallons
Engine Oil:	32 quarts

POWER TRAIN

Transfer Case: Planetary reduction gears
   Gear Ratio: 1.377:1 engine to transmission
Transmission: Torqmatic, 3 speeds forward, 1 reverse
   Torque Converter Ratio: Varies from 1:1 to 4.8:1

Gear Ratios:	1st	1:1	3rd	0.244:1
	2nd	0.428:1	reverse	0.756:1

Steering: Controlled differential
Bevel Gear Ratio: 3.53:1   Steering Ratio: 1.79:1
Brakes: Mechanical, 3 shoe, reverse anchor
Final Drive: Spur gear   Gear Ratio: 3.95:1
Drive Sprocket: At rear of vehicle with 13 teeth
   Pitch Diameter: 25.068 inches

RUNNING GEAR

Suspension: Torsion bar
   12 individually sprung dual road wheels (6/track)
   Tire Size: 26 x 6 inches
   10 dual track return rollers (5/track)
   Dual compensating idler at front of each track
   Idler Tire Size: 26 x 6 inches
   Shock absorbers fitted on first 2 and last 2 road wheels on each side
Tracks: Center guide, T81 and T80E1
   Type: (T81) Single pin, 24 inch width
         (T80E1) Double pin, 23 inch width
   Pitch: 6 inches
   Shoes per Vehicle: 164 (82/track)
   Ground Contact Length: 151.5 inches, left side
                          155.5 inches, right side

ELECTRICAL SYSTEM

Nominal Voltage: 24 volts DC
Generator: (1) 24 volts, 150 amperes, belt driven by either the main engine or the auxiliary engine
Battery: (2) 12 volts in series

COMMUNICATIONS

Radio: SCR 508 or 528 in rear of turret; SCR 506 (command tanks only) on shelf in front of loader; AN/VRC 3 in rear of turret on some tanks (infantry communication)
Interphone: (part of radio) 5 stations
Flag Set M238, Panel Set AP50A, Spotlight
Flares: 3 each, M17, M18, M19, and M21 (command tanks only)
Ground Signals Projector M4 (command tanks only)

FIRE AND GAS PROTECTION

(2) 10 pound carbon dioxide, fixed
(2) 4 pound carbon dioxide, portable
(2) 1½ quart decontaminating apparatus

PERFORMANCE

Maximum Speed: Sustained, level road		20 miles/hour
Short periods, level		25 miles/hour
Maximum Tractive Effort: TE at stall		60,000 pounds
Per Cent of Vehicle Weight: TE/W		62.5 per cent
Maximum Grade:		60 per cent
Maximum Trench:		8 feet
Maximum Vertical Wall:		46 inches
Maximum Fording Depth:		48 inches
Minimum Turning Circle: (diameter)		60 feet
Cruising Range: Roads	approx.	100 miles

MEDIUM TANK T26E5

GENERAL DATA

Crew:	5	men
Length: Gun forward, over pintle	340.5	inches
Length: Gun to rear, over sandshields	285	inches
Length: Over ss and pintle, w/o gun	249.1	inches
Gun Overhang: Gun forward	91.4	inches
Width: Over sandshields	148	inches
Height: To top of cupola	109.4	inches
Tread: With 23 inch tracks	110	inches
With 28 inch tracks	115	inches
Ground Clearance:	17.2	inches
Fire Height:	78	inches
Turret Ring Diameter: (inside)	69	inches
Weight, Combat Loaded:	102,300	pounds
Weight, Unstowed:	94,500	pounds
Power to Weight Ratio: Net	8.8	hp/ton
Gross	9.7	hp/ton
Ground Pressure: Zero penetration, 23" trk	14.5	psi
28" trk	11.9	psi

ARMOR

Type: Turret, cast homogeneous steel; Hull, rolled and cast homogeneous steel; Welded assembly

Hull Thickness:	Actual	Angle w/Vertical
Front, Upper	6.0 inches	46 degrees
Lower	4.0 inches	54 degrees
Sides, Front	3.0 inches	0 degrees
Rear	2.0 inches	0 degrees
Rear, Upper	2.0 inches	10 degrees
Lower	0.75 inches	62 degrees
Top	1.5 to 0.875 inches	90 degrees
Floor, Front	1.0 inches	90 degrees
Rear	0.5 inches	90 degrees
Turret Thickness:		
Gun Shield	11.0 inches	0 degrees
Front	7.5 inches	10 degrees
Sides	3.5 inches	0 to 8 degrees
Rear	5.0 inches	0 to 5 degrees
Top	1.0 inches	90 degrees

ARMAMENT

Primary: 90mm Gun M3 (T7) in Mount M67 (T99E2) in turret

Traverse: Hydraulic and manual	360 degrees
Traverse Rate: (max)	20 seconds/360 degrees
Elevation: Manual	+20 to -10 degrees
Firing Rate: (max)	8 rounds/minute
Loading System:	Manual
Stabilizer System:	None

Secondary:
(1) .50 caliber MG HB M2 flexible AA mount on turret
(1) .30 caliber MG M1919A4 coaxial w/90mm gun in turret
(1) .30 caliber MG M1919A4 in bow mount
Provision for (5) .45 caliber SMG M3
Provision for (1) .30 caliber Carbine M2 w/grenade launcher

AMMUNITION

70 rounds 90mm	12 hand grenades
550 rounds .50 caliber	
900 rounds .45 caliber	
5000 rounds .30 caliber	

FIRE CONTROL AND VISION EQUIPMENT*

Primary Weapon:	Direct	Indirect
	Telescope M83C	Azimuth Indicator M19
	Periscope M10F	Elevation Quadrant M9
Vision Devices:	Direct	Indirect
Driver	Hatch	Periscope M13 (1)
Asst. Driver	Hatch	Periscope M13 (1)
Commander	Vision cupola and hatch	Periscope M15 (1)
Gunner	None	Periscope M10F (1)
Loader	Hatch and pistol port	Periscope M13 (1)

Total Periscopes: M10F (1), M13 (3), M15 (1); In addition the following spares are carried, M10F (1), M13 (4)
Total Pistol Ports: Hull (0), Turret (1)
Vision Cupolas: (1) w/6 vision blocks on turret top

*Late production vision devices, drivers' auxiliary periscopes eliminated

ENGINE

Make and Model: Ford GAF	
Type: 8 cylinder, 4 cycle, 60 degree vee	
Cooling System: Liquid Ignition: Magneto	
Displacement:	1100 cubic inches
Bore and Stroke:	5.4 x 6 inches
Compression Ratio:	7.5:1
Net Horsepower (max):	450 hp at 2600 rpm
Gross Horsepower (max):	500 hp at 2600 rpm
Net Torque (max):	950 ft-lb at 2200 rpm
Gross Torque (max):	1040 ft-lb at 2200 rpm
Weight:	1414 lb, dry
Fuel: 80 octane gasoline	183 gallons
Engine Oil:	32 quarts

POWER TRAIN

Transfer Case: Planetary reduction gears
   Gear Ratio: 1.377:1 engine to transmission
Transmission: Torqmatic, 3 speeds forward, 1 reverse
   Torque Converter Ratio: Varies from 1:1 to 4.8:1

Gear Ratios:	1st	1:1	3rd	0.244:1
	2nd	0.428:1	reverse	0.756:1

Steering: Controlled differential
Bevel Gear Ratio: 3.53:1   Steering Ratio: 1.79:1
Brakes: Mechanical, 3 shoe, reverse anchor
Final Drive: Spur gear   Gear Ratio: 4.47:1
Drive Sprocket: At rear of vehicle with 13 teeth
   Pitch Diameter: 25.068 inches

RUNNING GEAR

Suspension: Torsion bar
   12 individually sprung dual road wheels (6/track)
   Tire Size: 26 x 6 inches
   10 dual track return rollers (5/track)
   Dual compensating idler at front of each track
   Idler Tire Size: 26 x 6 inches
   Shock absorbers fitted on first 2 and last 2 road wheels on each side
Tracks: Center guide, T80E1 w/5 inch end connectors
   Type: (T80E1) Double pin, 23 inch width
   Total Track Width: 28 inches with end connectors
   Pitch: 6 inches
   Shoes per Vehicle: 164 (82/track)
   Ground Contact Length: 151.5 inches, left side
                                      155.5 inches, right side

ELECTRICAL SYSTEM

Nominal Voltage: 24 volts DC
Generator: (1) 24 volts, 150 amperes, belt driven by either the main engine or the auxiliary engine
Battery: (2) 12 volts in series

COMMUNICATIONS

Radio: SCR 508 or 528 in rear of turret; SCR 506 (command tanks only) on shelf in front of loader; AN/VRC 3 in rear of turret on some tanks (infantry communication)
Interphone: (part of radio) 5 stations
Flag Set M238, Panel Set AP50A, Spotlight
Flares: 3 each, M17, M18, M19, and M21 (command tanks only)
Ground Signals Projector M4 (command tanks only)

FIRE AND GAS PROTECTION

(2) 10 pound carbon dioxide, fixed
(2) 4 pound carbon dioxide, portable
(2) 1½ quart decontaminating apparatus

PERFORMANCE

Maximum Speed: Sustained, level road		20 miles/hour
Short periods, level		25 miles/hour
Maximum Tractive Effort: TE at stall		68,000 pounds
Per Cent of Vehicle Weight: TE/W		66.5 per cent
Maximum Grade:		60 per cent
Maximum Trench:		8 feet
Maximum Vertical Wall:		46 inches
Maximum Fording Depth:		48 inches
Maximum Turning Circle: (diameter)		60 feet
Cruising Range: Roads	approx.	80 miles

MEDIUM TANK M26E1

GENERAL DATA
Crew: 5 men
Length: Gun forward, over pintle 376 inches
Length: Gun to rear, over sandshields 321 inches
Length: Over ss and pintle, w/o gun 249.1 inches
Gun Overhang: Gun forward 127 inches
Width: Over sandshields 138.3 inches
Height: To top of cupola 109.4 inches
Tread: 110 inches
Ground Clearance: 17.2 inches
Fire Height: 78 inches
Turret Ring Diameter: (inside) 69 inches
Weight, Combat Loaded: estimated 94,600 pounds
Weight, Unstowed: estimated 87,000 pounds
Power to Weight Ratio: Net 9.5 hp/ton
            Gross 10.6 hp/ton
Ground Pressure: Zero penetration 12.8 psi

ARMOR
Type: Turret, cast homogeneous steel; Hull, rolled and cast homogeneous steel; Welded assembly

Hull Thickness:

	Actual		Angle w/Vertical
Front, Upper	4.0	inches	46 degrees
Lower	3.0	inches	53 degrees
Sides, Front	3.0	inches	0 degrees
Rear	2.0	inches	0 degrees
Rear, Upper	2.0	inches	10 degrees
Lower	0.75	inches	62 degrees
Top	0.875	inches	90 degrees
Floor, Front	1.0	inches	90 degrees
Rear	0.5	inches	90 degrees

Turret Thickness:

	Actual		Angle w/Vertical
Gun Shield	4.5	inches	0 degrees
Front	4.0	inches	0 degrees
Sides	3.0	inches	0 to 8 degrees
Rear	3.0 inches plus counterweight		0 to 5 degrees
Top	1.0	inches	90 degrees

ARMAMENT
Primary: 90mm Gun T54 in Mount T126 in turret
Traverse: Hydraulic and manual 360 degrees
Traverse Rate: (max) 15 seconds/360 degrees
Elevation: Manual +15 to -10 degrees
Firing Rate: (max) 6 rounds/minute
Loading System: Manual
Stabilizer System: None
Secondary:
(1) .50 caliber MG HB M2 flexible AA mount on turret
(1) .50 caliber MG HB M2 coaxial w/90mm gun in turret
(1) .30 caliber MG M1919A4 in bow mount
Provision for (5) .45 caliber SMG M3

AMMUNITION
41 rounds 90mm 12 hand grenades
1100 rounds .50 caliber (estimated)
900 rounds .45 caliber
2500 rounds .30 caliber (estimated)

FIRE CONTROL AND VISION EQUIPMENT*

Primary Weapon:	Direct	Indirect
	Telescope M71E4	Azimuth Indicator M20
	Periscope M10E4	Elevation Quadrant M9
Vision Devices:	Direct	Indirect
Driver	Hatch	Periscope M13 (1)
Asst. Driver	Hatch	Periscope M13 (1)
Commander	Vision cupola and hatch	Periscope M15 (1)
Gunner	None	Periscope M10E4 (1)
Loader	Hatch and pistol port	Periscope M13 (1)

Total Periscopes: M10E4 (1), M13 (3), M15 (1); In addition the following spares are carried, M10E4 (1), M13 (4)
Total Pistol Ports: Hull (0), Turret (1)
Vision Cupolas: (1) w/6 vision blocks on turret top

*Late production vision devices, drivers' auxiliary periscopes eliminated

ENGINE
Make and Model: Ford GAF
Type: 8 cylinder, 4 cycle, 60 degree vee
Cooling System: Liquid   Ignition: Magneto
Displacement: 1100 cubic inches
Bore and Stroke: 5.4 x 6 inches
Compression Ratio: 7.5:1
Net Horsepower (max): 450 hp at 2600 rpm
Gross Horsepower (max): 500 hp at 2600 rpm
Net Torque (max): 950 ft-lb at 2200 rpm
Gross Torque (max): 1040 ft-lb at 2200 rpm
Weight: 1414 lb, dry
Fuel: 80 octane gasoline 183 gallons
Engine Oil: 32 quarts

POWER TRAIN
Transfer Case: Planetary reduction gears
  Gear Ratio: 1.377:1 engine to transmission
Transmission: Torqmatic, 3 speeds forward, 1 reverse
  Torque Converter Ratio: Varies from 1:1 to 4.8:1
  Gear Ratios: 1st   1:1   3rd   0.244:1
              2nd 0.428:1   reverse 0.756:1
Steering: Controlled differential
Bevel Gear Ratio: 3.53:1   Steering Ratio: 1.79:1
Brakes: Mechanical, 3 shoe, reverse anchor
Final Drive: Spur gear   Gear Ratio: 3.95:1
Drive Sprocket: At rear of vehicle with 13 teeth
  Pitch Diameter: 25.068 inches

RUNNING GEAR
Suspension: Torsion bar
  12 individually sprung dual road wheels (6/track)
  Tire Size: 26 x 6 inches
  10 dual track return rollers (5/track)
  Dual compensating idler at front of each track
  Idler Tire Size: 26 x 6 inches
  Shock absorbers fitted on first 2 and last 2 road wheels on each side
Tracks: Center guide, T81 and T80E1
  Type: (T81) Single pin, 24 inch width
       (T80E1) Double pin, 23 inch width
  Pitch: 6 inches
  Shoes per Vehicle: 164 (82/track)
  Ground Contact Length: 151.5 inches, left side
                    155.5 inches, right side

ELECTRICAL SYSTEM
Nominal Voltage: 24 volts DC
Generator: (1) 24 volts, 150 amperes, belt driven by either the main engine or the auxiliary engine
Battery: (2) 12 volts in series

COMMUNICATIONS
Radio: SCR 508 or 528 in rear of turret; SCR 506 (command tanks only) on shelf in front of loader; AN/VRC 3 in rear of turret on some tanks (infantry communication)
Interphone: (part of radio) 5 stations
Flag Set M238, Panel Set AP50A, Spotlight
Flares: 3 each, M17, M18, M19, and M21 (command tanks only)
Ground Signals Projector M4 (command tanks only)

FIRE AND GAS PROTECTION
(2) 10 pound carbon dioxide, fixed
(2) 4 pound carbon dioxide, portable
(2) 1½ quart decontaminating apparatus

PERFORMANCE
Maximum Speed: Sustained, level road 20 miles/hour
          Short periods, level 25 miles/hour
Maximum Tractive Effort: TE at stall 60,000 pounds
  Per Cent of Vehicle Weight: TE/W 63.4 per cent
Maximum Grade: 60 per cent
Maximum Trench: 8 feet
Maximum Vertical Wall: 46 inches
Maximum Fording Depth: 48 inches
Minimum Turning Circle: (diameter) 60 feet
Cruising Range: Roads approx. 100 miles

MEDIUM TANK M26E2

GENERAL DATA

Crew:	5 men
Length: Gun forward	335 inches
Length: Gun to rear	285 inches
Length: Without gun	244 inches
Gun Overhang: Gun forward	91 inches
Width: Over sandshields	138 inches
Height: To top of cupola	109 inches
Tread:	110 inches
Ground Clearance:	17 inches
Fire Height:	78 inches
Turret Ring Diameter: (inside)	69 inches
Weight, Combat Loaded:	estimated 94,000 pounds
Weight, Unstowed:	estimated 86,500 pounds
Power to Weight Ratio: Net	15.0 hp/ton
Gross	17.2 hp/ton
Ground Pressure: Zero penetration	13.3 psi

ARMOR

Type: Turret, cast homogeneous steel; Hull, rolled and cast homogeneous steel; Welded assembly

Hull Thickness:

		Actual	Angle w/Vertical
Front, Upper	4.0	inches	46 degrees
Lower	3.0	inches	53 degrees
Sides, Front	3.0	inches	0 degrees
Rear	2.0	inches	0 degrees
Rear, Upper	2.0	inches	10 degrees
Lower	0.75	inches	62 degrees
Top	0.875	inches	90 degrees
Floor, Front	1.0	inches	90 degrees
Rear	0.5	inches	90 degrees

Turret Thickness:

Gun Shield	4.5	inches	0 degrees
Front	4.0	inches	0 degrees
Sides	3.0	inches	0 to 8 degrees
Rear	3.0	inches	0 to 5 degrees
Top	1.0	inches	90 degrees

ARMAMENT

Primary: 90mm Gun M3 in Mount M67 in turret

Traverse: Hydraulic and Manual	360 degrees
Traverse Rate: (max)	15 seconds/360 degrees
Elevation: Manual	+20 to -10 degrees
Firing Rate: (max)	8 rounds/minute
Loading System:	Manual
Stabilizer System:	None

Secondary:
(1) .50 caliber MG HB M2 flexible AA mount on turret
(1) .30 caliber MG M1919A4 coaxial w/90mm gun in turret
(1) .30 caliber MG M1919A4 in bow mount
Provision for (5) .45 caliber SMG M3
Provision for (1) .30 caliber Carbine M2 w/grenade launcher

AMMUNITION

70 rounds 90mm	12 hand grenades
550 rounds .50 caliber	
900 rounds .45 caliber	
5000 rounds .30 caliber	

FIRE CONTROL AND VISION EQUIPMENT

Primary Weapon:	Direct	Indirect
	Telescope M71C	Azimuth Indicator M19
	Periscope M10F	Gunner's Quadrant M1
	Periscope M4A1, spare	Elevation Quadrant M9

Vision Devices:	Direct	Indirect
Driver	Hatch	Periscope M6 (2)
Asst. Driver	Hatch	Periscope M6 (2)
Commander	Vision cupola and hatch	Periscope M6 (1)
Gunner	None	Periscope M10F (1)
Loader	Hatch and pistol port	Periscope M6 (1)

Total Periscopes: M10F (1), M6 (6)
Total Pistol Ports: Hull (0), Turret (1)
Vision Cupolas: (1) w/6 vision blocks on turret top

ENGINE

Make and Model: Continental AV-1790-1 or -3	
Type: 12 cylinder, 4 cycle, 90 degree vee	
Cooling System: Air Ignition: -1 battery, -3 magneto	
Displacement:	1790 cubic inches
Bore and Stroke:	5.75 x 5.75 inches
Compression Ratio:	6.5:1
Net Horsepower (max):	704 hp at 2800 rpm
Gross Horsepower (max):	810 hp at 2800 rpm
Net Torque (max):	1440 ft-lb at 2200 rpm
Gross Torque (max):	1610 ft-lb at 2200 rpm
Weight:	2380 lb, dry
Fuel: 80 octane gasoline	190 gallons
Engine Oil:	72 quarts

POWER TRAIN

Transmission: Cross-drive, 2 ranges forward, 1 reverse
   Single stage hydraulic converter
   Stall Multiplication: 5.2:1
   Overall Usable Ratios: low 15.5   reverse 20.6
                          high 7.1
Steering Control: Mechanical wobble stick
   Steering Rate: 5.7 RPM
Brakes: Multiple disc
Final Drive: Spur gear   Gear Ratio: 3.95:1
Drive Sprocket: At rear of vehicle with 13 teeth
   Pitch Diameter: 25.068 inches

RUNNING GEAR

Suspension: Torsion bar
   12 individually sprung dual road wheels (6/track)
   Tire Size: 26 x 6 inches
   10 dual track return rollers (5/track)
   Dual compensating idler at front of each track
   Idler Tire Size: 26 x 6 inches
   Shock absorbers fitted on first 2 and last 2 road wheels on each side
Tracks: Center guide, T81 and T80E1
   Type: (T81) Single pin, 24 inch width
        (T80E1) Double pin, 23 inch width
   Pitch: 6 inches
   Shoes per Vehicle: 164 (82/track)
   Ground Contact Length: 151.5 inches, left side
                          155.5 inches, right side

ELECTRICAL SYSTEM

Nominal Voltage: 24 volts DC
Main Generator: (1) 24 volts, 175 amperes, driven by power take-off from main engine
Auxiliary Generator: (1) 24 volts, 175 amperes, driven by the auxiliary engine
Battery: (4) 12 volts

COMMUNICATIONS

Radio: SCR 508 or 528 in rear of turret; SCR 506 (command tanks only) on shelf in front of loader; AN/VRC 3 in rear of turret on some tanks (infantry communication)
Interphone: (part of radio) 5 stations
Flag Set M238, Panel Set AP50A, Spotlight
Flares: 3 each, M17, M18, M19, and M21 (command tanks only)
Ground Signals Projector M4 (command tanks only)

FIRE AND GAS PROTECTION

(2) 10 pound carbon dioxide, fixed
(2) 4 pound carbon dioxide, portable
(2) 1½ quart decontaminating apparatus

PERFORMANCE

Maximum Speed: Sustained, level road	25 miles/hour
Maximum Tractive Effort: TE at stall	81,000 pounds
Per Cent of Vehicle Weight: TE/W	86.2 per cent
Maximum Grade:	60 per cent
Maximum Trench:	8 feet
Maximum Vertical Wall:	46 inches
Maximum Fording Depth:	48 inches
Minimum Turning Circle: (diameter)	pivot
Cruising Range: Roads	175 miles

8 inch HOWITZER MOTOR CARRIAGE T84

GENERAL DATA

Crew:		8	men
Length: Overall		273	inches
Width: Over sandshields		136.3	inches
Height: Overall		125	inches
Tread:		110	inches
Ground Clearance:		17.7	inches
Weight, Combat Loaded:	estimated	82,000	pounds
Weight, Unstowed:		76,660	pounds
Power to Weight Ratio: Net		11.0	hp/ton
Gross		12.2	hp/ton
Ground Pressure: Zero penetration		11.8	psi

ARMOR

Type: Rolled homogeneous steel, welded assembly

Hull Thickness:	Actual	Angle w/Vertical
Front, Upper	1.0 inches	64 degrees
Lower	1.0 inches	35 degrees
Sides, Upper	0.5 inches	0 degrees
Lower	1.0 inches	0 degrees
Rear	0.5 inches	30 degrees
Top	0.5 inches	90 degrees
Floor, Front	1.0 inches	90 degrees
Rear	0.5 inches	90 degrees
Howitzer Shield	0.5 inches	40 degrees

ARMAMENT

Primary: 8 inch Howitzer M1 in mount at rear of hull

Traverse: Manual	15 degrees each side
Elevation: Manual	+60 to -5 degrees
Firing Rate: (max)	1 round/minute
Loading System:	Manual

Secondary:
  Provision for (1) .30 caliber Rifle M1903A3
  Provision for (7) .30 caliber Carbines M2
  Provision for (1) Grenade Launcher M8

AMMUNITION

6 rounds 8 inch	12 hand grenades
10 rifle grenades	

FIRE CONTROL AND VISION EQUIPMENT

Primary Weapon: Telescope M69E2
  Panoramic Telescope M12
  Gunner's Quadrant M1

Vision Devices:	Direct	Indirect
Driver	Vision cupola and hatch	None
Asst. Driver	Vision cupola and hatch	None

Vision Cupolas: (2) w/4 vision blocks on hull roof

ENGINE

Make and Model: Ford GAF	
Type: 8 cylinder, 4 cycle, 60 degree vee	
Cooling System: Liquid Ignition: Magneto	
Displacement:	1100 cubic inches
Bore and Stroke:	5.4 x 6 inches
Compression Ratio:	7.5:1
Net Horsepower (max):	450 hp at 2600 rpm
Gross Horsepower (max):	500 hp at 2600 rpm
Net Torque (max):	950 ft-lb at 2200 rpm
Gross Torque (max):	1040 ft-lb at 2200 rpm
Weight:	1414 lb, dry
Fuel: 80 octane gasoline	200 gallons
Engine Oil:	32 quarts

POWER TRAIN

Transfer Case: Planetary reduction gears
  Gear Ratio: 1.377:1 engine to transmission
Transmission: Torqmatic, 3 speeds forward, 1 reverse
  Torque Converter Ratio: Varies from 1:1 to 4.8:1

Gear Ratios:	1st	1:1	3rd	0.244:1
	2nd	0.428:1	reverse	0.756:1

Steering: Controlled differential
  Bevel Gear Ratio: 3.53:1   Steering Ratio: 1.79:1
Brakes: Mechanical, 3 shoe, reverse anchor
Final Drive: Spur gear   Gear Ratio: 3.95:1
Drive Sprocket: At front of vehicle with 13 teeth
  Pitch Diameter: 25.068 inches

RUNNING GEAR

Suspension: Torsion bar
  12 individually sprung dual road wheels (6/track)
  Tire Size: 26 x 6 inches
  10 dual track return rollers (5/track)
  Dual compensating idler at rear of each track
  Idler Tire Size: 26 x 6 inches
  Shock absorbers fitted on first 2 and last 2 road wheels on each side
Tracks: Center guide, T81 and T80E1
  Type: (T81) Single pin, 24 inch width
    (T80E1) Double pin, 23 inch width
  Pitch: 6 inches
  Shoes per Vehicle: 164 (82/track)
  Ground Contact Length: 149.6 inches, right side
    153.4 inches, left side

ELECTRICAL SYSTEM

Nominal Voltage: 24 volts DC
Generator: (1) 24 volts, 75 amperes, driven by power take-off from main engine
Battery: (2) 12 volts in series

COMMUNICATIONS

Radio: SCR 610 or 628 in left sponson compartment
Interphone: RC 99, 4 stations

FIRE AND GAS PROTECTION

(2) 10 pound carbon dioxide, fixed
(2) 4 pound carbon dioxide, portable
(1) 1½ quart decontaminating apparatus

PERFORMANCE

Maximum Speed: Sustained, level road		25 miles/hour
Maximum Tractive Effort: TE at stall		60,000 pounds
Per Cent of Vehicle Weight: TE/W		73.2 per cent
Maximum Grade:		60 per cent
Maximum Trench:		8 feet
Maximum Vertical Wall:		36 inches
Maximum Fording Depth:		36 inches
Minimum Turning Circle: (diameter)		60 feet
Cruising Range: Roads	approx.	100 miles

8 inch GUN MOTOR CARRIAGE T93

GENERAL DATA

Crew:	8	men
Length: Overall	447	inches
Width: With 23 inch tracks, w/o ss	133	inches
With 28 inch tracks, w/o ss	143	inches
Height: Overall	126.9	inches
Tread: With 23 inch tracks	110	inches
With 28 inch tracks	115	inches
Ground Clearance:	18.7	inches
Weight, Combat Loaded:	132,600	pounds
Weight, Unstowed: estimated	128,000	pounds
Power to Weight Ratio: Net	6.8	hp/ton
Gross	7.5	hp/ton
Ground Pressure: Zero penetration, 23" trk	16.0	psi
28" trk	13.2	psi

ARMOR

Type: Rolled homogeneous steel, welded assembly

Hull Thickness:	Actual		Angle w/Vertical
Front, Upper	1.0	inches	64 to 72 degrees
Lower	1.0	inches	40 degrees
Sides, Upper	0.5	inches	0 degrees
Lower	1.0	inches	0 degrees
Rear	0.5	inches	10 degrees
Top	0.875	inches	90 degrees
Floor	1.0	inches	90 degrees
Gun Shield	0.5	inches	30 degrees

ARMAMENT

Primary: 8 inch Gun M1 in Mount T31 at rear of hull

Traverse: Manual	12 degrees each side
Elevation: Manual	+65 to 0 degrees
Firing Rate: (max)	1 round/minute
Loading System:	Manual

Secondary:
Provision for (1) .30 caliber Rifle M1903A3
Provision for (7) .30 caliber Carbines M2
Provision for (1) Grenade Launcher M8

AMMUNITION

10 rifle grenades   12 hand grenades

FIRE CONTROL AND VISION EQUIPMENT

Primary Weapon: Elbow Telescope M16A1E1
                   Panoramic Telescope M12
                   Gunner's Quadrant M1

Vision Devices:	Direct	Indirect
Driver	Vision cupola and hatch	None
Asst. Driver	Vision cupola and hatch	None

Vision Cupolas: (2) w/3 vision blocks on front hull plate

ENGINE

Make and Model: Ford GAF	
Type: 8 cylinder, 4 cycle, 60 degree vee	
Cooling System: Liquid Ignition: Magneto	
Displacement:	1100 cubic inches
Bore and Stroke:	5.4 x 6 inches
Compression Ratio:	7.5:1
Net Horsepower (max):	450 hp at 2600 rpm
Gross Horsepower (max):	500 hp at 2600 rpm
Net Torque (max):	950 ft-lb at 2200 rpm
Gross Torque (max):	1040 ft-lb at 2200 rpm
Weight:	1414 lb, dry
Fuel: 80 octane gasoline	245 gallons
Engine Oil:	32 quarts

POWER TRAIN

Transfer Case: Planetary reduction gears
    Gear Ratio: 1.377:1 engine to transmission
Transmission: Torqmatic, 3 speeds forward, 1 reverse
    Torque Converter Ratio: Varies from 1:1 to 4.8:1

Gear Ratios:	1st	1:1	3rd	0.244:1
	2nd	0.428:1	reverse	0.756:1

Steering: Controlled differential
    Bevel Gear Ratio: 3.53:1   Steering Ratio: 1.79:1
Brakes: Mechanical, 3 shoe, reverse anchor
Final Drive: Planetary gear   Gear Ratio: 6.25:1
Drive Sprocket: At front of vehicle with 13 teeth
    Pitch Diameter: 25.068 inches

RUNNING GEAR

Suspension: Torsion bar
    14 individually sprung dual road wheels (7/track)
    Tire Size: 26 x 6 inches
    12 dual track return rollers (6/track)
    Dual compensating idler at rear of each track
    Idler Tire Size: 26 x 6 inches
    Shock absorbers fitted on first 2 and last 2 road wheels on each side
Tracks: Center guide, T80E1 w/5 inch end connectors
    Type: (T80E1) Double pin, 23 inch width
    Total Track Width: 28 inches with end connectors
    Pitch: 6 inches
    Shoes per Vehicle: 188 (94/track)
    Ground Contact Length: 180 inches, average

ELECTRICAL SYSTEM

Nominal Voltage: 24 volts DC
Main Generator: (1) 24 volts, 50 amperes, driven by power take-off from main engine
Auxiliary Generator: (1) 24 volts, 50 amperes, driven by the auxiliary engine
Battery: (2) 12 volts in series

COMMUNICATIONS

Interphone: RC 99, 4 stations
Remote Control Equipment: RC 261

FIRE AND GAS PROTECTION

(2) 10 pound carbon dioxide, fixed
(2) 4 pound carbon dioxide, portable
(2) 1½ quart decontaminating apparatus

PERFORMANCE

Maximum Speed: Sustained, level road		15 miles/hour
Short periods, level		20 miles/hour
Maximum Tractive Effort: TE at stall		95,000 pounds
Per Cent of Vehicle Weight: TE/W		71.6 per cent
Maximum Grade:		60 per cent
Maximum Trench:		8 feet
Maximum Vertical Wall:		46 inches
Maximum Fording Depth:		36 inches
Minimum Turning Circle: (diameter)		70 feet
Cruising Range: Roads	approx.	50 miles

240mm HOWITZER MOTOR CARRIAGE T92

GENERAL DATA

Crew:	8	men
Length: Overall	382	inches
Width: With 23 inch tracks, w/o ss	133	inches
With 28 inch tracks, w/o ss	143	inches
Height: Overall	126.9	inches
Tread: With 23 inch tracks	110	inches
With 28 inch tracks	115	inches
Ground Clearance:	18.7	inches
Weight, Combat Loaded:	127,500	pounds
Weight, Unstowed: estimated	123,000	pounds
Power to Weight Ratio: Net	7.1	hp/ton
Gross	7.8	hp/ton
Ground Pressure: Zero penetration, 23" trk	15.4	psi
28" trk	12.6	psi

ARMOR

Type: Rolled homogeneous steel, welded assembly

Hull Thickness:	Actual		Angle w/Vertical
Front, Upper	1.0	inches	64 to 72 degrees
Lower	1.0	inches	40 degrees
Sides, Upper	0.5	inches	0 degrees
Lower	1.0	inches	0 degrees
Rear	0.5	inches	10 degrees
Top	0.875	inches	90 degrees
Floor	1.0	inches	90 degrees
Howitzer Shield	0.5	inches	30 degrees

ARMAMENT

Primary: 240mm Howitzer M1 in Mount T30 at rear of hull

Traverse: Manual	12 degrees each side
Elevation: Manual	+65 to 0 degrees
Firing Rate: (max)	1 round/minute
Loading System:	Manual

Secondary:
Provision for (1) .30 caliber Rifle M1903A3
Provision for (7) .30 caliber Carbines M2
Provision for (1) Grenade Launcher M8

AMMUNITION

10 rifle grenades   12 hand grenades

FIRE CONTROL AND VISION EQUIPMENT

Primary Weapon: Elbow Telescope M16A1E1
Panoramic Telescope M12
Gunner's Quadrant M1

Vision Devices:	Direct	Indirect
Driver	Vision cupola and hatch	None
Asst. Driver	Vision cupola and hatch	None

Vision Cupolas: (2) w/3 vision blocks on front hull plate

ENGINE

Make and Model: Ford GAF	
Type: 8 cylinder, 4 cycle, 60 degree vee	
Cooling System: Liquid Ignition: Magneto	
Displacement:	1100 cubic inches
Bore and Stroke:	5.4 x 6 inches
Compression Ratio:	7.5:1
Net Horsepower (max):	450 hp at 2600 rpm
Gross Horsepower (max):	500 hp at 2600 rpm
Net Torque (max):	950 ft-lb at 2200 rpm
Gross Torque (max):	1040 ft-lb at 2200 rpm
Weight:	1414 lb, dry
Fuel: 80 octane gasoline	245 gallons
Engine Oil:	32 quarts

POWER TRAIN

Transfer Case: Planetary reduction gears
Gear Ratio: 1.377:1 engine to transmission
Transmission: Torqmatic, 3 speeds forward, 1 reverse
Torque Converter Ratio: Varies from 1:1 to 4.8:1

Gear Ratios:	1st	1:1	3rd	0.244:1
	2nd	0.428:1	reverse	0.756:1

Steering: Controlled differential
Bevel Gear Ratio: 3.53:1   Steering Ratio: 1.79:1
Brakes: Mechanical, 3 shoe, reverse anchor
Final Drive: Planetary gear   Gear Ratio: 6.25:1
Drive Sprocket: At front of vehicle with 13 teeth
Pitch Diameter: 25.068 inches

RUNNING GEAR

Suspension: Torsion bar
14 individually sprung dual road wheels (7/track)
Tire Size: 26 x 6 inches
12 dual track return rollers (6/track)
Dual compensating idler at rear of each track
Idler Tire Size: 26 x 6 inches
Shock absorbers fitted on first 2 and last 2 road wheels on each side
Tracks: Center guide, T80E1 w/5 inch end connectors
Type: (T80E1) Double pin, 23 inch width
Total Track Width: 28 inches with end connectors
Pitch: 6 inches
Shoes per Vehicle: 188 (94/track)
Ground Contact Length: 180 inches, average

ELECTRICAL SYSTEM

Nominal Voltage: 24 volts DC
Main Generator: (1) 24 volts, 50 amperes, driven by power take-off from main engine
Auxiliary Generator: (1) 24 volts, 50 amperes, driven by the auxiliary engine
Battery: (2) 12 volts in series

COMMUNICATIONS

Interphone: RC 99, 4 stations
Remote Control Equipment: RC 261

FIRE AND GAS PROTECTION

(2) 10 pound carbon dioxide, fixed
(2) 4 pound carbon dioxide, portable
(2) 1½ quart decontaminating apparatus

PERFORMANCE

Maximum Speed: Sustained, level road		15 miles/hour
Short periods, level		20 miles/hour
Maximum Tractive Effort: TE at stall		95,000 pounds
Per Cent of Vehicle Weight: TE/W		74.5 per cent
Maximum Grade:		60 per cent
Maximum Trench:		8 feet
Maximum Vertical Wall:		46 inches
Maximum Fording Depth:		36 inches
Minimum Turning Circle: (diameter)		70 feet
Cruising Range: Roads	approx.	50 miles

CARGO CARRIER T31

GENERAL DATA

Crew: 6 men
Length: Overall 252 inches
Width: Over sandshields 136.3 inches
Height: Overall 106 inches
Tread: 110 inches
Ground Clearance: 17.7 inches
Weight, Combat Loaded: estimated 80,000 pounds
Weight, Unstowed: estimated 50,000 pounds
Power to Weight Ratio: Net 11.3 hp/ton
                Gross 12.5 hp/ton
Ground Pressure: Zero penetration 11.5 psi

ARMOR

Type: Rolled homogeneous steel, welded assembly

Hull Thickness:	Actual	Angle w/Vertical
Front, Upper	1.0 inches	64 degrees
Lower	1.0 inches	35 degrees
Sides, Upper	0.5 inches	0 degrees
Lower	1.0 inches	0 degrees
Rear, Upper	0.5 inches	0 degrees
Lower	0.5 inches	30 degrees
Top	0.5 inches	90 degrees
Floor, Front	1.0 inches	90 degrees
Rear	0.5 inches	90 degrees

ARMAMENT

(1) .50 caliber MG HB M2 on ring mt. over asst. driver
Provision for (1) .30 caliber Rifle M1903A3
Provision for (5) .30 caliber Carbines M2
Provision for (1) Grenade Launcher M8

AMMUNITION*

60 rounds 8 inch howitzer   12 hand grenades
600 rounds .50 caliber
10 rifle grenades

VISION EQUIPMENT

	Direct	Indirect
Driver	Vision cupola and hatch	None
Asst. Driver	Vision cupola and hatch	None

Vision Cupolas: (2) w/4 vision blocks on hull roof

*Alternate stowage arrangements for ammunition for the 240mm howitzer, 8 inch gun, 155mm gun, or 155mm howitzer

ENGINE

Make and Model: Ford GAF
Type: 8 cylinder, 4 cycle, 60 degree vee
Cooling System: Liquid   Ignition: Magneto
Displacement: 1100 cubic inches
Bore and Stroke: 5.4 x 6 inches
Compression Ratio: 7.5:1
Net Horsepower (max): 450 hp at 2600 rpm
Gross Horsepower (max): 500 hp at 2600 rpm
Net Torque (max): 950 ft-lb at 2200 rpm
Gross Torque (max): 1040 ft-lb at 2200 rpm
Weight: 1414 lb, dry
Fuel: 80 octane gasoline 200 gallons
Engine Oil: 32 quarts

POWER TRAIN

Transfer Case: Planetary reduction gears
   Gear Ratio: 1.377:1 engine to transmission
Transmission: Torqmatic, 3 speeds forward, 1 reverse
   Torque Converter Ratio: Varies from 1:1 to 4.8:1

Gear Ratios:	1st	1:1	3rd	0.244:1
	2nd	0.428:1	reverse	0.756:1

Steering: Controlled differential
   Bevel Gear Ratio: 3.53:1   Steering Ratio: 1.79:1
Brakes: Mechanical, 3 shoe, reverse anchor
Final Drive: Spur gear   Gear Ratio: 3.95:1
Drive Sprocket: At front of vehicle with 13 teeth
   Pitch Diameter: 25.068 inches

RUNNING GEAR

Suspension: Torsion bar
   12 individually sprung dual road wheels (6/track)
   Tire Size: 26 x 6 inches
   10 dual track return rollers (5/track)
   Dual compensating idler at rear of each track
   Idler Tire Size: 26 x 6 inches
   Shock absorbers fitted on first 2 and last 2 road wheels on each side
Tracks: Center guide, T81 and T80E1
   Type: (T81) Single pin, 24 inch width
         (T80E1) Double pin, 23 inch width
   Pitch: 6 inches
   Shoes per Vehicle: 164 (82/track)
   Ground Contact Length: 149.6 inches, right side
                   153.4 inches, left side

ELECTRICAL SYSTEM

Nominal Voltage: 24 volts DC
Generator: (1) 24 volts, 75 amperes, driven by power take-off from main engine
Battery: (2) 12 volts in series

COMMUNICATIONS

Radio: SCR 610 or 628 in left sponson compartment
Interphone: RC 99, 4 stations

FIRE AND GAS PROTECTION

(2) 10 pound carbon dioxide, fixed
(2) 4 pound carbon dioxide, portable
(1) 1½ quart decontaminating apparatus

PERFORMANCE

Maximum Speed: Sustained, level road 25 miles/hour
Maximum Tractive Effort: TE at stall 60,000 pounds
   Per Cent of Vehicle Weight: TE/W 75.0 per cent
Maximum Grade: 60 per cent
Maximum Trench: 8 feet
Maximum Vertical Wall: 36 inches
Maximum Fording Depth: 36 inches
Minimum Turning Circle: (diameter) 60 feet
Cruising Range: Roads approx. 100 miles

TANK RECOVERY VEHICLE T12

GENERAL DATA
Crew: 6 men
Length: Boom to rear, travel position   298 inches
Length: Without boom   263 inches
Boom Overhang: Boom to rear   35 inches
Width: Over sandshields   148 inches
Height: Boom in travel position   134 inches
Tread: With 23 inch tracks   110 inches
      With 28 inch tracks   115 inches
Ground Clearance: 18 inches
Weight, Combat Loaded: estimated   90,000 pounds
Weight, Unstowed: estimated   80,000 pounds
Power to Weight Ratio: Net   10.0 hp/ton
               Gross   11.1 hp/ton
Ground Pressure: Zero penetration, 23" trk   12.7 psi
                           28" trk   10.5 psi

ARMOR
Type: Turret, rolled homogeneous steel; Hull, rolled and cast homogeneous steel; Welded assembly

Hull Thickness:

		Actual		Angle w/Vertical
Front,	Upper	4.0	inches	46 degrees
	Lower	3.0	inches	53 degrees
Sides,	Front	3.0	inches	0 degrees
	Rear	2.0	inches	0 degrees
Rear,	Upper	2.0	inches	10 degrees
	Lower	0.75	inches	62 degrees
Top		0.875	inches	90 degrees
Floor,	Front	1.0	inches	90 degrees
	Rear	0.5	inches	90 degrees

Turret Thickness:

		Actual		Angle w/Vertical
Front		1.25	inches	0 degrees
Sides		1.25	inches	0 degrees
Rear		1.25	inches	0 degrees
Top		1.0	inches	90 degrees

ARMAMENT
(1) .50 caliber MG HB M2 on ring mount on turret top
(1) .30 caliber MG M1919A4 in bow mount
Provision for (2) .45 caliber SMG M3
Provision for (4) .30 caliber Carbines M2
Provision for (1) 2.36 inch Rocket Launcher M9A1
Provision for (1) Grenade Launcher M8

AMMUNITION
550 rounds .50 caliber   10 rockets 2.36 inch
360 rounds .45 caliber   10 rifle grenades
2000 rounds .30 caliber   12 hand grenades

VISION EQUIPMENT

	Direct	Indirect
Driver	Hatch	Periscope M6 (2)
Asst. Driver	Hatch	Periscope M6 (2)
Commander	Vision cupola and hatch	Periscope M6 (1)

Total Periscopes: M6 (5)
Vision Cupolas: (1) w/6 vision blocks on turret top

ENGINE
Make and Model: Ford GAF
Type: 8 cylinder, 4 cycle, 60 degree vee
Cooling System: Liquid   Ignition: Magneto
Displacement: 1100 cubic inches
Bore and Stroke: 5.4 x 6 inches
Compression Ratio: 7.5:1
Net Horsepower (max):   450 hp at 2600 rpm
Gross Horsepower (max):   500 hp at 2600 rpm
Net Torque (max):   950 ft-lb at 2200 rpm
Gross Torque (max):   1040 ft-lb at 2200 rpm
Weight: 1414 lb, dry
Fuel: 80 octane gasoline   202 gallons
Engine Oil: 32 quarts

POWER TRAIN
Transfer Case: Planetary reduction gears
      Gear Ratio: 1.377:1 engine to transmission
Transmission: Torqmatic, 3 speeds forward, 1 reverse
      Torque Converter Ratio: Varies from 1:1 to 4.8:1
      Gear Ratios: 1st   1:1   3rd   0.244:1
                     2nd   0.428:1   reverse   0.756:1
Steering: Controlled differential
      Bevel Gear Ratio: 3.53:1   Steering Ratio: 1.79:1
Brakes: Mechanical, 3 shoe, reverse anchor
Final Drive: Spur gear   Gear Ratio: 3.95:1
Drive Sprocket: At rear of vehicle with 13 teeth
      Pitch Diameter: 25.068 inches

RUNNING GEAR
Suspension: Torsion bar
      12 individually sprung dual road wheels (6/track)
      Tire Size: 26 x 6 inches
      10 dual track return rollers (5/track)
      Dual compensating idler at front of each track
      Idler Tire Size: 26 x 6 inches
      Shock absorbers fitted on first 2 and last 2 road wheels on each side
Tracks: Center guide, T80E1 w/5 inch end connectors
      Type: (T80E1) Double pin, 23 inch width
      Total Track Width: 28 inches with end connectors
      Pitch: 6 inches
      Shoes per Vehicle: 164 (82/track)
      Ground Contact Length: 151.5 inches, left side
                                    155.5 inches, right side

ELECTRICAL SYSTEM
Nominal Voltage: 24 volts DC
Generator: (1) 24 volts, 150 amperes, belt driven by either the main engine or the auxiliary engine
Battery: (2) 12 volts in series

COMMUNICATIONS
Radio: SCR 508 or 528 on shelf in turret
Interphone: (part of radio) 5 stations plus RC 298 external interphone extension
Flag Set M238, Panel Set AP50A, Spotlight

FIRE AND GAS PROTECTION
      (2) 10 pound carbon dioxide, fixed
      (2) 4 pound carbon dioxide, portable
      (2) 1 quart carbon tetrachloride, portable
      (1) 1½ quart decontaminating apparatus

PERFORMANCE
Maximum Speed: Sustained, level road   25 miles/hour
                     Short periods, level   30 miles/hour
Maximum Tractive Effort: TE at stall   60,000 pounds
      Per Cent of Vehicle Weight: TE/W   66.7 per cent
Maximum Grade: 60 per cent
Maximum Trench: 8 feet
Maximum Vertical Wall: 46 inches
Maximum Fording Depth: 48 inches
Minimum Turning Circle: (diameter)   60 feet
Cruising Range: Roads   approx.   100 miles

The general characteristics and performance data for the major armament mounted on the T20 series vehicles are included in the following pages. To avoid confusion, caliber lengths are quoted for each weapon using both the U.S. (bore length only) and the German method (muzzle to rear face of breech). Other dimensions are defined in the sketch below.

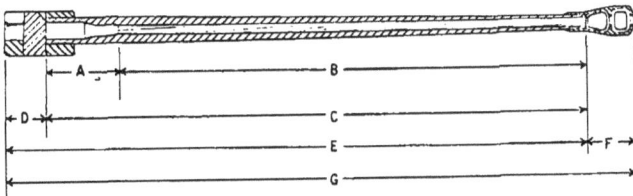

A. Length of Chamber (to rifling)
B. Length of Rifling
C. Length of Bore
D. Depth of Breech Recess
E. Length, Muzzle to Rear Face of Breech
F. Additional Length, Muzzle Brake, Etc.
G. Overall Length

Official nomenclature in effect during the period of greatest service is used to describe each type of ammunition. Since the terminalogy changed occasionally, a standard nomenclature is added in parenthesis. An explanation of the standard terms, which are used separately and in combination, is given below:

AP	Armor piercing, uncapped
APBC	Armor piercing with ballistic cap, without armor piercing cap
APCBC	Armor piercing with armor piercing cap and ballistic cap
APCR	Armor piercing, composite rigid
CP	Concrete piercing
HE	High explosive
HEAT	High explosive antitank, hollow charge
-T	Tracer

All armor plate angles quoted in the data sheets are measured between a vertical plane and the plate surface. This is illustrated by angle α in the figure below. The armor plate penetration performance of the various projectiles is usually quoted for a 30 degree angle of obliquity. This angle is defined as the angle between a line perpendicular to the armor plate and the projectile path. This is shown by angle β below.

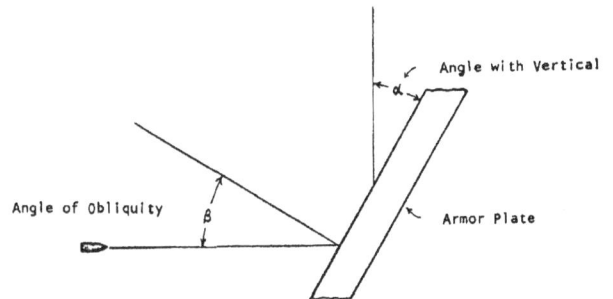

In regard to armor penetration, it should be noted that the values obtained at 30 degrees obliquity do not necessarily indicate the relative performance of these projectiles at other angles. For example, the 90 mm AP T33 is approximately equal to the APC M82 at 30 degrees, but at the 55 degree angle of the Panther's front plate, the latter fails completely while the T33 will penetrate at ranges up to 1100 yards. Against this type of high obliquity target, the mono-bloc shot was also much more effective than the HVAP or APCR type projectile.

75mm Gun M3

Carriage and Mount	Medium Tank M3 series in Mount M1, Medium Tank M4 series in Mounts M34 and M34A1, Medium Tank M7 in Mount M47, Medium Tank T22E1 in Mount M34 with an automatic loader, Assault Tank T14 in Mount M34A1
Length of Chamber (to rifling)	14.4 inches
Length of Rifling	96.2 inches
Length of Chamber (to projectile base)	12.96 inches (APC M61), 11.5 inches (HE M48)
Travel of Projectile in Bore	97.67 inches (APC M61), 99.1 inches (HE M48)
Length of Bore	110.63 inches, 37.5 calibers
Depth of Breech Recess	7.75 inches
Length, Muzzle to Rear Face of Breech	118.38 inches, 40.1 calibers
Additional Length, Muzzle Brake, etc.	None
Overall Length	118.38 inches
Diameter of Bore	2.95 inches
Chamber Capacity	88.05 cubic inches (APC M61), 80.57 cubic inches (HE M48)
Weight, Tube	611 pounds
Total Weight	893 pounds
Type of Breechblock	Semiautomatic sliding wedge. Gun mounted so breechblock slides vertically in Mount M1, at 45 degrees in Mount M47, and horizontally in Mounts M34 and M34A1
Rifling	24 grooves, uniform right-hand twist, one turn in 25.59 calibers (slope 7 degrees)
Ammunition	Fixed
Primer	Percussion
Weight, Complete Round	APC M61 Projectile (APCBC/HE-T) 19.92 pounds
	AP M72 Shot (AP-T) 18.80 pounds
	HE M48 Shell (HE), Supercharge 19.56 pounds
	HE M48 Shell (HE), Normal 18.80 pounds
	HC BI M89 Shell, Smoke 9.83 pounds
Weight, Projectile	APC M61 Projectile (APCBC/HE-T) 14.96 pounds
	AP M72 Shot (AP-T) 13.94 pounds
	HE M48 Shell (HE) 14.70 pounds
	HC BI M89 Shell, Smoke 6.61 pounds
Maximum Powder Pressure	38,000 psi
Maximum Rate of Fire	20 rounds/minute
Muzzle Velocity	APC M61 Projectile (APCBC/HE-T) 2,030 ft/sec
	AP M72 Shot (AP-T) 2,030 ft/sec
	HE M48 Shell (HE), Supercharge 1,980 ft/sec
	HE M48 Shell (HE), Normal 1,520 ft/sec
	HC BI M89 Shell, Smoke 850 ft/sec
Muzzle Energy of Projectile, KE=½MV^2 Rotational energy is neglected and values are based on long tons (2,240 pounds)	APC M61 Projectile (APCBC/HE-T) 427 ft-tons
	AP M72 Shot (AP-T) 398 ft-tons
	HE M48 Shell (HE), Supercharge 400 ft-tons
	HE M48 Shell (HE), Normal 235 ft-tons
Maximum Range (independent of mount)	APC M61 Projectile (APCBC/HE-T) 14,000 yards
	AP M72 Shot (AP-T) 10,650 yards
	HE M48 Shell (HE), Supercharge 14,000 yards
	HE M48 Shell (HE), Normal 11,400 yards
	HC BI M89 Shell, Smoke approx. 1,500 yards

Penetration Performance

Homogeneous Armor at 30 degrees obliquity

Range	500 yards	1,000 yards	1,500 yards	2,000 yards
APC M61	2.6 inches (66mm)	2.4 inches (60mm)	2.2 inches (55mm)	2.0 inches (50mm)
AP M72	3.0 inches (76mm)	2.5 inches (63mm)	2.0 inches (51mm)	1.7 inches (43mm)

Face-Hardened Armor at 30 degrees obliquity

Range	500 yards	1,000 yards	1,500 yards	2,000 yards
APC M61	2.9 inches (74mm)	2.6 inches (67mm)	2.4 inches (60mm)	2.1 inches (54mm)
AP M72	2.6 inches (66mm)	2.1 inches (53mm)	1.6 inches (41mm)	1.3 inches (33mm)

76mm Gun M1, M1A1, M1A1C, and M1A2

Carriage and Mount	Medium Tank M4 series in Mount M62, Medium Tanks T20, T20E3, T22, and T23-2 in Mount T79, Medium Tanks T23 and T23E3 in Mount T80, Motor Carriages M18, T86, and T86E1 in Mount M1
Length of Chamber (to rifling)	22.46 inches
Length of Rifling	133.54 inches
Length of Chamber (to projectile base)	20.7 inches (square base projectiles)
Travel of Projectile in Bore	135.3 inches (square base projectiles)
Length of Bore	156.00 inches, 52.0 calibers
Depth of Breech Recess	7.75 inches
Length, Muzzle to Rear Face of Breech	163.75 inches, 54.6 calibers
Additional Length, Muzzle Brake, etc.	11.6 inches, Muzzle Brake M2 on M1A1C and M1A2
Overall Length	163.75 inches (M1, M1A1), 175.4 inches (M1A1C, M1A2)
Diameter of Bore	3.000 inches
Chamber Capacity	142.6 cubic inches (APC M62), 140.50 inches (HE M42A1)
Weight, Tube (without muzzle brake)	870 pounds (M1), 940 pounds (M1A1)
Weight, Complete (without muzzle brake)	1,141 pounds (M1), 1,206 pounds (M1A1C), 1,231 pounds (M1A2)
Weight, Muzzle Brake M2	62 pounds
Total Weight	1,141 pounds (M1), 1,268 pounds (M1A1C), 1,293 pounds (M1A2)
Type of Breechblock	Semiautomatic sliding wedge. Gun mounted so breechblock slides horizontally in Mounts M62, T79, and T80, and at 45 degrees in Mount M1
Rifling	28 grooves, uniform right-hand twist, one turn in 40 calibers (M1, M1A1, M1A1C) or 32 calibers (M1A2)
Ammunition	Fixed
Primer	Percussion
Weight, Complete Round	APC M62 Projectile (APCBC/HE-T) 24.80 pounds
	HVAP M93 Shot (APCR-T) 18.91 pounds
	AP M79 Shot (AP-T) 24.24 pounds
	HE M42A1 Shell (HE) 22.23 pounds
	HC BI M88 Shell, Smoke 13.43 pounds
Weight, Projectile	APC M62 Projectile (APCBC/HE-T) 15.44 pounds
	HVAP M93 Shot (APCR-T) 9.40 pounds
	AP M79 Shot (AP-T) 15.00 pounds
	HE M42A1 Shell (HE) 12.87 pounds
	HC BI M88 Shell, Smoke 7.38 pounds
Maximum Powder Pressure	43,000 psi
Maximum Rate of Fire	20 rounds/minute
Muzzle Velocity	APC M62 Projectile (APCBC/HE-T) 2,600 ft/sec
	HVAP M93 Shot (APCR-T) 3,400 ft/sec
	AP M79 Shot (AP-T) 2,600 ft/sec
	HE M42A1 Shell (HE) 2,700 ft/sec
	HC BI M88 Shell, Smoke 900 ft/sec
Muzzle Energy of Projectile, $KE = \frac{1}{2}MV^2$ Rotational energy is neglected and values are based on long tons (2,240 pounds)	APC M62 Projectile (APCBC/HE-T) 724 ft-tons
	HVAP M93 Shot (APCR-T) 753 ft-tons
	AP M79 Shot (AP-T) 703 ft-tons
	HE M42A1 Shell (HE) 650 ft-tons
Maximum Range (independent of mount)	APC M62 Projectile (APCBC/HE-T) 16,100 yards
	HVAP M93 Shot (APCR-T) 13,100 yards
	AP M79 Shot (AP-T) 12,770 yards
	HE M42A1 Shell (HE) 14,200 yards
	HC BI M88 Shell, Smoke (at 12 degrees) 2,000 yards

Penetration Performance — Homogeneous Armor at 30 degrees obliquity

Range	500 yards	1,000 yards	1,500 yards	2,000 yards
APC M62	3.7 inches (93mm)	3.5 inches (88mm)	3.2 inches (82mm)	3.0 inches (75mm)
HVAP M93	6.2 inches (157mm)	5.3 inches (135mm)	4.6 inches (116mm)	3.9 inches (98mm)
AP M79	4.3 inches (109mm)	3.6 inches (92mm)	3.0 inches (76mm)	2.5 inches (64mm)

Penetration values are for the M1A1 gun. A slight increase for APC is obtained with the M1A2 at the longer ranges

229

90mm Gun M3

Carriage and Mount	Medium Tanks T25 and T25E1 in Mount T99, Medium Tanks T26 and T26E1 in Mount T99E1, Medium Tanks M26 (T26E3), M26E2, and T26E5 in Mount M67 (T99E2), Motor Carriage M36 series in Mount M4 (T8). This weapon mounted experimentally in the Heavy Tank T1E1 and Motor Carriages M10 and M18
Length of Chamber (to rifling)	24.8 inches
Length of Rifling	152.4 inches
Length of Chamber (to projectile base)	20.8 inches (boat-tailed projectiles)
Travel of Projectile in Bore	156.4 inches (boat-tailed projectiles)
Length of Bore	177.15 inches, 50.0 calibers
Depth of Breech Recess	9.00 inches
Length, Muzzle to Rear Face of Breech	186.15 inches, 52.5 calibers
Additional Length, Muzzle Brake, etc.	16.0 inches, Muzzle Brake M3 on late production
Overall Length	202.2 inches with muzzle brake
Diameter of Bore	3.543 inches
Chamber Capacity	300 cubic inches
Weight, Complete (without muzzle brake)	2,300 pounds
Weight, Muzzle Brake M3	149.5 pounds
Total Weight	2,450 pounds
Type of Breechblock	Semiautomatic, vertical sliding wedge
Rifling	32 grooves, uniform right-hand twist, one turn in 32 calibers
Ammunition	Fixed
Primer	Percussion
Weight, Complete Round	APC M82 Projectile (APCBC/HE-T) early 42.75 pounds
	APC M82 Projectile (APCBC/HE-T) late 43.87 pounds
	HVAP T30E16 Shot (APCR-T) 37.13 pounds
	AP T33 Shot (APBC-T) 43.82 pounds
	HE M71 Shell (HE) 41.93 pounds
Weight, Projectile	APC M82 Projectile (APCBC/HE-T) 24.11 pounds
	HVAP M304 (T30E16) Shot (APCR-T) 16.80 pounds
	AP T33 Shot (APBC-T) 24.06 pounds
	HE M71 Shell (HE) 23.29 pounds
Maximum Powder Pressure	38,000 psi
Maximum Rate of Fire	8 rounds/minute
Muzzle Velocity	APC M82 Projectile (APCBC/HE-T) early 2,650 ft/sec
	APC M82 Projectile (APCBC/HE-T) late 2,800 ft/sec
	HVAP M304 (T30E16) Shot (APCR-T) 3,350 ft/sec
	AP T33 Shot (APBC-T) 2,800 ft/sec
	HE M71 Shell (HE) 2,700 ft/sec
Muzzle Energy of Projectile, $KE = \frac{1}{2}MV^2$ Rotational energy is neglected and values are based on long tons (2,240 pounds)	APC M82 Projectile (APCBC/HE-T) early 1,174 ft-tons
	APC M82 Projectile (APCBC/HE-T) late 1,310 ft-tons
	HVAP M304 (T30E16) Shot (APCR-T) 1,307 ft-tons
	AP T33 Shot (APBC-T) 1,310 ft-tons
	HE M71 Shell (HE) 1,177 ft-tons
Maximum Range (independent of mount)	APC M82 Projectile (APCBC/HE-T) early 20,400 yards
	APC M82 Projectile (APCBC/HE-T) late 21,400 yards
	HVAP M304 (T30E16) Shot (APCR-T) 15,700 yards
	AP T33 Shot (APBC-T) 21,000 yards
	HE M71 Shell (HE) 19,560 yards

Penetration Performance

Homogeneous Armor at 30 degrees obliquity

Range	500 yards	1,000 yards	1,500 yards	2,000 yards
APC M82 (2,650)	4.7 inches (120mm)	4.4 inches (112mm)	4.1 inches (104mm)	3.8 inches (96mm)
APC M82 (2,800)	5.1 inches (129mm)	4.8 inches (122mm)	4.5 inches (114mm)	4.2 inches (106mm)
HVAP M304 (T30E16)	8.7 inches (221 mm)	7.9 inches (199mm)	7.0 inches (176mm)	6.1 inches (156mm)
AP T33	4.7 inches (119mm)	4.6 inches (117mm)	4.5 inches (114mm)	4.3 inches (109mm)

90mm Gun T15E2

Carriage and Mount	Medium Tank T26E4 in Mount T119, Heavy Tanks T32 and T32E1 in Mount T119
Length of Chamber (to rifling)	40.7 inches
Length of Rifling	207.2 inches
Length of Chamber (to projectile base)	36.7 inches (boat-tailed projectiles)
Travel of Projectile in Bore	211.2 inches (boat-tailed projectiles)
Length of Bore	247.9 inches, 70.0 calibers
Depth of Breech Recess	9.0 inches
Length, Muzzle to Rear Face of Breech	256.9 inches, 72.6 calibers
Additional Length, Muzzle Brake, etc.	16.0 inches, Muzzle Brake M3
Overall Length	272.9 inches
Diameter of Bore	3.543 inches
Chamber Capacity	488 cubic inches (estimated)
Weight, Complete (without muzzle brake)	3,270 pounds
Weight, Muzzle Brake M3	150 pounds
Total Weight	3,420 pounds
Type of Breechblock	Semiautomatic, vertical sliding wedge
Rifling	32 grooves, uniform right-hand twist, one turn in 32 calibers
Ammunition	Separated
Primer	Percussion

Weight, Complete Round	AP T43 Shot (APBC-T)	51.2 pounds
	HVAP T44 Shot (APCR-T)	44 pounds
	HE T42 Shell (HE)	50.4 pounds
Weight, Projectile	AP T43 Shot (APBC-T)	24.06 pounds
	HVAP T44 Shot (APCR-T)	16.70 pounds
	HE T42 Shell (HE)	23.3 pounds
Maximum Powder Pressure	41,500 psi	
Maximum Rate of Fire	4 rounds/minute	
Muzzle Velocity	AP T43 Shot (APBC-T)	3,200 ft/sec
	HVAP T44 Shot (APCR-T)	3,750 ft/sec
	HE T42 Shell (HE)	3,200 ft/sec
Muzzle Energy of Projectile, $KE=\frac{1}{2}MV^2$	AP T43 Shot (APBC-T)	1,711 ft-tons
Rotational energy is neglected and	HVAP T44 Shot (APCR-T)	1,628 ft-tons
values are based on long tons (2,240 pounds)	HE T42 Shell (HE)	1,654 ft-tons
Maximum Range (independent of mount)	HE T42 Shell (HE)	27,000 yards

Penetration Performance   Homogeneous Armor at 30 degrees obliquity

Range	500 yards	1,000 yards	1,500 yards	2,000 yards
AP T43	5.2 inches (132mm)	5.0 inches (127mm)	4.9 inches (124mm)	4.8 inches (122mm)
HVAP T44	9.6 inches (244mm)	8.7 inches (221mm)	7.7 inches (196mm)	6.8 inches (173mm)

90mm Gun T54

Carriage and Mount	Medium Tank M26E1 in Mount T126
Length of Chamber (to rifling)	24.6 inches
Length of Rifling	209.184 inches
Length of Chamber (to projectile base)	21.8 inches
Travel of Projectile in Bore	212 inches
Length of Bore	233.8 inches, 66.6 calibers
Depth of Breech Recess	8.45 inches
Length, Muzzle to Rear Face of Breech	242.25 inches, 68.4 calibers
Additional Length, Muzzle Brake, etc.	6.75 inches, Muzzle Brake M3E2
Overall Length	249.0 inches
Diameter of Bore	3.543 inches
Chamber Capacity	465 cubic inches
Weight, Complete (without muzzle brake)	3,400 pounds
Weight, Muzzle Brake	80 pounds
Total Weight	3,480 pounds
Type of Breechblock	Semiautomatic, vertical sliding wedge
Rifling	32 grooves, uniform right-hand twist, one turn in 32 calibers
Ammunition	Fixed
Primer	Percussion
Weight, Complete Round	AP Shot (APBC-T)	approx. 49 pounds
	HVAP Shot (APCR-T)	approx. 42 pounds
	HE Shell (HE)	approx. 48 pounds
Weight, Projectile	AP Shot (APBC-T)	24.1 pounds
	HVAP Shot (APCR-T)	16.7 pounds
	HE Shell (HE)	23.3 pounds
Maximum Powder Pressure	40,000 psi
Maximum Rate of Fire	6 rounds/minute
Muzzle Velocity	AP Shot (APBC-T)	3,200 ft/sec
	HVAP Shot (APCR-T)	3,750 ft/sec
	HE Shell (HE)	3,200 ft/sec
Muzzle Energy of Projectile, $KE = \frac{1}{2}MV^2$ Rotational energy is neglected and values are based on long tons (2,240 pounds)	AP Shot (APBC-T)	1,711 ft-tons
	HVAP Shot (APCR-T)	1,628 ft-tons
	HE Shell (HE)	1,654 ft-tons
Maximum Range (independent of mount)	HE Shell (HE)	27,000 yards

Penetration Performance — Homogeneous Armor at 30 degrees obliquity

Range	500 yards	1,000 yards	1,500 yards	2,000 yards
AP	5.2 inches (132mm)	5.0 inches (127mm)	4.9 inches (124mm)	4.8 inches (122mm)
HVAP	9.6 inches (244mm)	8.7 inches (221mm)	7.7 inches (196mm)	6.8 inches (173mm)

105mm Howitzer M4

Carriage and Mount	Medium Tank M4 series in Mount M52, Medium Tank M45 (T26E2) in Mount M71 (T117), Motor Carriage M37 in Mount M5
Length of Chamber (to rifling)	15.03 inches
Length of Rifling	78.02 inches
Length of Chamber (to projectile base)	11.38 inches (boat-tailed projectiles)
Travel of Projectile in Bore	81.67 inches (boat-tailed projectiles)
Length of Bore	93.05 inches, 22.5 calibers
Depth of Breech Recess	8.25 inches
Length, Muzzle to Rear Face of Breech	101.3 inches, 24.5 calibers
Additional Length, Muzzle Brake, etc.	None
Overall Length	101.3 inches
Diameter of Bore	4.134 inches
Chamber Capacity	153.80 cubic inches
Weight, Tube	973 pounds
Total Weight	1,140 pounds
Type of Breechblock	Manually operated, horizontal sliding wedge
Rifling	36 grooves, uniform right-hand twist, one turn in 20 calibers
Ammunition	Semifixed, variable charge except for HEAT M67
Primer	Percussion
Weight, Complete Round	HE M1 Shell (HE) 42.07 pounds
	HEAT M67 Shell (HEAT-T) 36.85 pounds
	WP M60 Shell, Smoke 43.77 pounds
	HC BE M84 Shell, Smoke 41.94 pounds
Weight, Projectile	HE M1 Shell (HE) 33.00 pounds
	HEAT M67 Shell (HEAT-T) 29.22 pounds
	WP M60 Shell, Smoke 34.31 pounds
	HC BE M84 Shell, Smoke 32.87 pounds
Maximum Powder Pressure	28,000 psi
Maximum Rate of Fire	8 rounds/minute
Muzzle Velocity	HE M1 Shell (HE) maximum charge 1,550 ft/sec
	HEAT M67 Shell (HEAT-T) 1,250 ft/sec
	WP M60 Shell, Smoke maximum charge 1,550 ft/sec
	HC BE M84 Shell, Smoke maximum charge 1,550 ft/sec
Muzzle Energy of Projectile, $KE=\frac{1}{2}MV^2$ Rotational energy is neglected and values are based on long tons (2,240 pounds)	HE M1 Shell (HE) maximum charge 555 ft-tons
	HEAT M67 Shell (HEAT-T) 317 ft-tons
Maximum Range (independent of mount)	HE M1 Shell (HE) maximum charge 12,205 yards
	HEAT M67 Shell (HEAT-T) 8,590 yards
	WP M60 Shell, Smoke maximum charge 12,150 yards
	HC BE M84 Shell, Smoke maximum charge 12,205 yards
Penetration Performance	Homogeneous Armor at 0 degrees obliquity, HEAT M67, 4.0 inches at any range

8 inch Howitzer M1 and M2

Carriage and Mount	Towed Carriage M1, Motor Carriage T84, Motor Carriage M43 in Mount M17
Length of Chamber (to rifling)	35.2 inches
Length of Rifling	164.8 inches
Length of Chamber (to projectile base)	26.2 inches (HE M106)
Travel of Projectile in Bore	173.83 inches (HE M106)
Length of Bore	200.0 inches, 25.0 calibers
Depth of Breech Recess	9.6 inches
Length, Muzzle to Rear Face of Breech	209.59 inches, 26.2 calibers
Additional Length, Breech Mechanism	6.4 inches
Overall Length	216.0 inches
Diameter of Bore	8.00 inches
Chamber Capacity	1,527 cubic inches (HE M106)
Total Weight	10,240 pounds
Type of Breechblock	Stepped thread interrupted screw, horizontal swing
Rifling	64 grooves, uniform right-hand twist, one turn in 25 calibers
Ammunition	Separate Loading
Primer	Percussion
Weight, Complete Round	HE M106 Shell (HE) with Charge M2 — 228.19 pounds
	HE Mk1A1 Shell (HE) with Charge M1 — 213.19 pounds
Weight, Projectile	HE M106 Shell (HE) — 200.00 pounds
	HE Mk1A1 Shell (HE) — 200.00 pounds
Maximum Powder Pressure	33,000 psi
Maximum Rate of Fire	1 round/minute
Muzzle Velocity	HE M106 Shell (HE) with Charge M2 — 1,950 ft/sec
	HE Mk1A1 Shell (HE) with Charge M1 — 1,339 ft/sec
Muzzle Energy of Projectile, $KE=\frac{1}{2}MV^2$ Rotational energey is neglected and values are based on long tons (2,240 pounds)	HE M106 Shell (HE) with Charge M2 — 5,272 ft-tons
Maximum Range (independent of mount)	HE M106 Shell (HE) with Charge M2 — 18,510 yards
	HE Mk1A1 Shell (HE) with Charge M1 — 11,170 yards
Penetration Performance	Reinforced Concrete at 0 degrees obliquity, HE M106 with CP Fuse M78; 5.5 feet at 1,000 yards, 4.7 feet at 3,000 yards, 4.0 feet at 5,000 yards, 3.2 feet at 10,000 yards, 3.1 feet at 15,000 yards

8 inch Gun M1

Carriage and Mount	Towed Carriage M2, Motor Carriage T93 in Mount T31
Length of Chamber (to rifling)	71.8 inches
Length of Rifling	328.15 inches
Length of Chamber (to projectile base)	61.4 inches (HE M103)
Travel of Projectile in Bore	338.58 inches (HE M103)
Length of Bore	400.0 inches, 50.0 calibers
Depth of Breech Recess	9.5 inches
Length, Muzzle to Rear Face of Breech	409.5 inches, 51.2 calibers
Additional Length, Breech Mechanism	5.0 inches (estimated)
Overall Length	415 inches (estimated)
Diameter of Bore	8.00 inches
Chamber Capacity	5,156 cubic inches (HE M103)
Total Weight	29,800 pounds
Type of Breechblock	Stepped thread interrupted screw, vertical swing
Rifling	64 grooves, uniform right-hand twist, one turn in 25 calibers
Ammunition	Separate Loading
Primer	Percussion
Weight, Complete Round	HE M103 Shell (HE) maximum charge 346 pounds
	Mk14 Projectile, Common (APBC/HE) max. chg. 366 pounds
Weight, Projectile	HE M103 Shell (HE) 240 pounds
	Mk14 Projectile, Common (APBC/HE) 260 pounds
Maximum Powder Pressure	38,000 psi
Maximum Rate of Fire	1 round/minute
Muzzle Velocity	HE M103 Shell (HE) maximum charge 2,850 ft/sec
	Mk14 Projectile, Common (APBC/HE) max. chg. 2,750 ft/sec
Muzzle Energy of Projectile, KE=½MV^2 Rotational energy is neglected and values are based on long tons (2,240 pounds)	HE M103 Shell (HE) maximum charge 13,514 ft-tons
Maximum Range (independent of mount)	HE M103 Shell (HE) maximum charge 35,630 yards
	Mk14 Projectile, Common (APBC/HE) max. chg. 32,500 yards
Penetration Performance	Reinforced Concrete at 0 degrees obliquity, HE M103 with CP Fuse M78; 5.2 feet at 15,000 yards, 4.5 feet at 19,000 yards, 3.8 feet at 23,000 yards, 3.7 feet at 27,000 yards

240mm Howitzer M1

Carriage and Mount	Towed Carriage M1, Motor Carriage T92 in Mount T30	
Length of Chamber (to rifling)	63.8 inches	
Length of Rifling	257.69 inches	
Length of Chamber (to projectile base)	53.3 inches (HE M114)	
Travel of Projectile in Bore	268.16 inches (HE M114)	
Length of Bore	321.5 inches, 34.0 calibers	
Depth of Breech Recess	9.5 inches	
Length, Muzzle to Rear Face of Breech	331.0 inches, 35.0 calibers	
Additional Length, Breech Mechanism	5.0 inches (estimated)	
Overall Length	336 inches (estimated)	
Diameter of Bore	9.45 inches	
Chamber Capacity	4,430.0 cubic inches (HE M114)	
Total Weight	25,100 pounds	
Type of Breechblock	Stepped thread interrupted screw, vertical swing	
Rifling	68 grooves, uniform right-hand twist, one turn in 25 calibers	
Ammunition	Separate Loading	
Primer	Percussion	
Weight, Complete Round	HE M114 Shell (HE) maximum charge	438.78 pounds
Weight, Projectile	HE M114 Shell (HE)	360 pounds
Maximum Powder Pressure	36,000 psi	
Maximum Rate of Fire	1 round/minute	
Muzzle Velocity	HE M114 Shell (HE) maximum charge	2,300 ft/sec
Muzzle Energy of Projectile, KE=½MV^2	HE M114 Shell (HE) maximum charge	13,202 ft-tons
Rotational energy is neglected and values are based on long tons (2,240 pounds)		
Maximum Range (independent of mount)	HE M114 Shell (HE) maximum charge	25,255 yards
Penetration Performance	Reinforced Concrete at 0 degrees obliquity, HE M114 with CP Fuse M78; 4.9 feet at 12,500 yards, 4.4 feet at 15,000 yards, 4.1 feet at 19,000 yards	

REFERENCES AND SELECTED BIBLIOGRAPHY

Books and Manuscripts

Hechler, Ken, "The Bridge at Remagen," Ballantine Books, New York, New York, 1957

————, "Spearhead in the West," *The History of the Third Armored Division 1941-1945*

Green, Constance McL., Harry C. Thompson, and Peter C. Roots, "The Ordnance Department: Planning Munitions for War," *The U.S. Army in World War II*, Office of the Chief of Military History, Washington D.C., 1955

Thomson, Harry C. and Lida Mayo, "The Ordnance Department: Procurement and Supply," *The U.S. Army in World War II*, Office of the Chief of Military History, Washington D.C., 1960

Mayo, Lida, "The Ordnance Department on Beachhead and Battlefront," *The U.S. Army in World War II*, Office of the Chief of Military History, Washington D.C., 1968

Chase, Daniel, "A History of Combat Vehicle Development," unpublished manuscript

Whiting, T. E., "Statistics," Manuscript prepared for the *History of the U.S. Army in World War II*, Office of the Chief of Military History, Washington D.C., 9 April 1952

Appleman, Roy E., "South to the Naktong, North to the Yalu," *The U.S. Army in the Korean War*, Office of the Chief of Military History, Washington D.C., 1961

Reports and Official Documents

"Summary of Combat Vehicle Development," Office Chief of Ordnance, Detroit, 1945

"The Design, Development, and Production of Tanks in World War II," Office Chief of Ordnance, Washington D.C., 15 August 1944

"Record of Army Ordnance Research and Development—Tanks," OCO, Washington D.C., undated

"Catalogue of Standard Ordnance Items—Tanks and Automotive Vehicles," OCO, Washington D.C., Revised to June 1945

"A Handbook of Ordnance Automotive Engineering—Volume 1, Combat and Tracklaying Vehicles," Aberdeen Proving Ground, Maryland, 1 May 1945

"Report of the Fiscal Year 1944-1945, Research and Development Service," OCO, Washington D.C., October 1945

"New Ordnance Materiel," OCO, Washington D.C., January 1945

"Pacific Area Materiel," OCO, Washington D.C., July 1945

"Notes on Materiel, Medium Tanks T20 and T20E3," Fisher Body Division, General Motors Corporation, Detroit Michigan, undated

"Notes on Materiel, Medium Tank T22," Chrysler Corporation Detroit, Michigan, undated

"First Report on Automatic 75 mm gun in Medium Tank T22E1 and First Report on Ordnance Program No. 5738," APG, Maryland, 2 April 1945

"Preliminary Specification RY-24184, T23 Medium Tank built by Erie Works," General Electric Company, Schenectady, N.Y., March 1943

"Final report on the Ten Production Models Medium Tank T23, Project No. P-417-1," The Armored Board, Fort Knox, Kentucky, 4 November 1944

"Pilot Tank T23E3 with Torsion Bar Suspension, Report No. T-61505.38-01," Chrysler Engineering Division, Detroit, Michigan, 14 November 1944

"Final Report of Test of Three Medium Tanks T23 with 23 inch track and horizontal volute spring suspension, Report No. P659," The Armored Board, Fort Knox, Kentucky, 27 August 1945

"Technical Manual TM9-734, Medium Tank T23," War Department, Washington D.C., 24 March 1944

"Pilot Tanks, T25 Medium Tanks Equipped with 90 mm Gun, Report No. T-61505-01," Chrysler Engineering Division, Detroit, Michigan, 23 June 1944

"Final Report of Test of Medium Tank T25, Project No. P-595," The Armored Board, Fort Knox, Kentucky, 14 August 1945

"Final Report of Test of Medium Tank T25E1, Project No. P-545," The Armored Board, Fort Knox, Kentucky, 17 March 1945

"Notes on Materiel, Medium Tanks T25E1 and T26E1," War Department, Washington D.C., June 1944

"Notes on Materiel, 90 mm Combination Gun Mount T102," Rock Island Arsenal, June 1945

"T26 Heavy Tank Pilot Vehicle, Report No. T-61905.8," Chrysler Engineering Division, Detroit, Michigan, 4 May 1945

"Final Report on Comparative Test of Heavy Tank T26 and Heavy Tank T26E1, Project No. P-636," The Armored Board, Fort Knox, Kentucky, 28 September 1945

"First Report on Test of Heavy Tank T26E1 and First Report on Ordnance Program 6004," APG, Maryland, 26 May 1944

"Final Report on Test of Heavy Tank T26E3, Project No. P-546-1," The Armored Board, Fort Knox, Kentucky, 9 May 1945

"Technical Manual TM9-735, Heavy Tank T26E3," War Department, Washington D.C., 15 January 1945

"Technical Manual TM9-735, Medium Tanks M26 and M45," War Department, Washington D.C., August 1948

"Technical Manual TM9-1735A, Medium Tanks M26 and M45, Power Trains," War Department, Washington D.C., September 1947

"Technical Manual TM9-1735B, Medium Tanks M26 and M45, Tracks, Suspension, Hull, and Turret," War Department, Washington D.C., September 1947

"First Report on Medium Tank T26E4 and Seventeenth Report on Ordnance Program 6009," APG, Maryland, 14 February 1947

"First Report on Test of Medium Tank T26E5 and Twelfth Report on Ordnance Program No. 6009," APG, Maryland, 22 July 1946

"Test of 90 mm Combination Gun Mount T126 in Medium Tank M26E1, First Report on Project TT2-6-34," APG, Maryland, 15 April 1949

"First Partial Report on Pilot Model Test of Medium Tank M26E2 and Medium Tank T40 and Second Report on Project TT2-668," APG, Maryland, 27 June 1950

"Notes on Materiel, Multiple Rocket Launcher, 4.5 inch T99," Chrysler Corporation, Detroit, Michigan, 15 November 1945

"Notes on Materiel, Eight Inch Howitzer Motor Carriage T84," Chrysler Engineering Division, Detroit, Michigan, undated

"Notes on Materiel, Eight Inch Gun Motor Carriage T93," Chrysler Corporation, Detroit, Michigan, 15 August 1945

"Notes on Materiel, 240 mm Howitzer Motor Carriage T92," Chrysler Corporation, Detroit, Michigan, 15 July 1945

"Notes on Materiel, Eight Inch Gun Motor Carriage T93," Chrysler Corporation, Detroit, Michigan, 15 August 1945

"Notes on Materiel, 240 mm Howitzer Motor Carriage T92," Chrysler Corporation, Detroit, Michigan, 15 July 1945

"Test of 240 mm Howitzer Motor Carriage T92 and First Report on Project TT2-538," APG, Maryland, 3 May 1949

"Notes on Materiel, Cargo Carrier T31," Chrysler Engineering Division, Detroit, Michigan, undated

"Notes on Materiel, Tank Recovery Vehicle T12," Pressed Steel Car Company, Inc., Chicago, Illinois, 1 December 1945

"Three M26 tanks at Chinju 31 July 1950," Inspector General Report, 10 September 1950

"The Employment of Armor in Korea, Vol. I," The Armor School, Fort Knox, Kentucky 1952

"Summary of Activities," 2nd Battalion 27th Infantry, 21-22nd August 1950

"Technical Manual TM9-1901, Artillery Ammunition," War Department, Washington D.C., 29 June 1944 and Department of the Army, September 1950

"Technical Manual TM9-1907, Ballistic Performance Data of Ammunition," Department of the Army, Washington D.C., July 1948

INDEX